VOICES FROM THE MOON

VOICES FROM THE MOON

APOLLO ASTRONAUTS DESCRIBE THEIR LUNAR EXPERIENCES

Andrew Chaikin with Victoria Kohl

VIKING STUDIO
Published by the Penguin Group
Penguin Group (USA) Inc., 375 Hudson Street, New York, New York 10014, U.S.A. •
Penguin Group (Canada), 90 Eglinton Avenue East, Suite 700, Toronto, Ontario,
Canada M4P 2Y3 (a division of Pearson Penguin Canada Inc.) • Penguin Books Ltd,
80 Strand, London WC2R 0RL, England • Penguin Ireland, 25 St. Stephen's Green,
Dublin 2, Ireland (a division of Penguin Books Ltd) • Penguin Books Australia Ltd, 250
Camberwell Road, Camberwell, Victoria 3124, Australia (a division of Pearson Austra-
lia Group Pty Ltd) • Penguin Books India Pvt Ltd, 11 Community Centre, Panchsheel
Park, New Delhi – 110 017, India • Penguin Group (NZ), 67 Apollo Drive, Rosedale,
North Shore 0632, New Zealand (a division of Pearson New Zealand Ltd) • Penguin
Books (South Africa) (Pty) Ltd, 24 Sturdee Avenue, Rosebank, Johannesburg 2196,
South Africa

Penguin Books Ltd, Registered Offices:
80 Strand, London WC2R 0RL, England

First published in 2009 by Viking Studio,
a member of Penguin Group (USA) Inc.

10 9 8 7 6 5 4 3 2 1

Copyright © Andrew Chaikin, 2009
All rights reserved

Photograph credits appear on page 201.

No copyright is claimed on works of the United States Government.

978-0-670-02078-2

Printed in China
Set in Avenir

Designed by Daniel Lagin
Photo layout by Andrew Chaikin

This book is dedicated to the four hundred thousand individuals—
the men and women of Apollo—who worked to turn a science fiction
dream into reality. Without them, the words and images on these
pages would not exist.

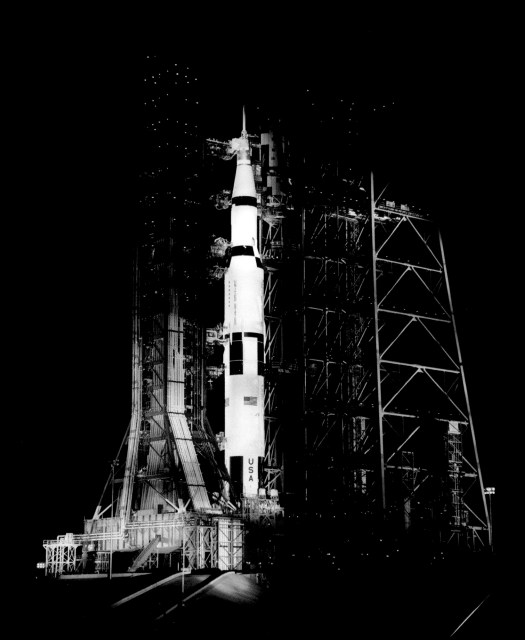

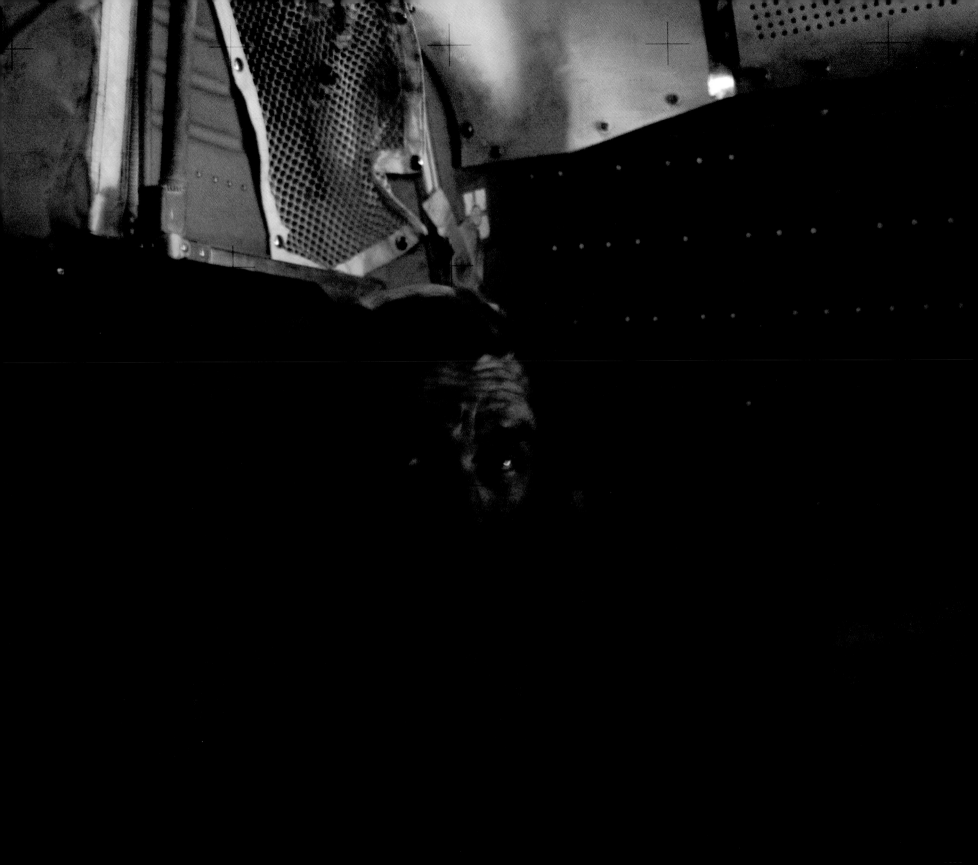

CONTENTS

INTRODUCTION

"I CANNOT COUNT THE NUMBER OF TIMES SOMEBODY HAS said to me, 'What does it feel like to be on the moon?' I'm tired of that, Andy. I don't want to try and figure that out anymore. I've done my best."

I could hear the exasperation in Dave Scott's voice. On a Saturday evening in August 1987, at a restaurant near Los Angeles, I was several hours into the second of two marathon interviews with the commander of Apollo 15, conversations that would eventually total almost twelve hours. I was in the third year of what would become an eight-year effort to interview twenty-three of the twenty-four lunar astronauts (Apollo 13's Jack Swigert had died in 1982) to tell the stories of their missions in my book, *A Man on the Moon: The Voyages of the Apollo Astronauts*. In truth, I had been preparing for these interviews my entire life.

I was a child of the space age, obsessed with space exploration for as long as I could remember, and I'd devoured every book on the subject I could get my hands on. After I saw Gemini astronauts walking in space I began "training" for it myself, anytime I was in a swimming pool. During every Apollo flight I camped out in front of the TV with maps of the moon and models of the spacecraft, my own little mission control in the den. As a teenager

I badgered my parents to take me to NASA headquarters and to Cape Canaveral, where I got to meet some of my heroes in person. I even wrote to my congressman for a VIP pass to the night launch of Apollo 17, the final moon mission, in 1972. In hopes of becoming a scientist-astronaut, I studied planetary geology and took part in the first Mars landing, Viking, as an undergraduate intern. But a couple of years after I graduated, I changed course. Instead of trying to be a space explorer, I became a space journalist. By 1984 I'd zeroed in on the mission that would take me as close to the moon as I could get: to document the Apollo astronauts' lunar experiences.

By the time I began my interviews for the book in June 1985 with Apollo 12 commander Pete Conrad, in his office at the McDonnell Douglas Corporation in Long Beach, California, I'd read every Apollo 12 transcript and debriefing, listened to all the on-board voice tapes, and watched every one of the mission's onboard TV transmissions. It didn't take too long for Conrad to say to me, "You know this mission better than I do." What I hadn't yet real-ized was that I was under the influence of my own expectations. I was *sure* anyone who went to the moon would undergo a kind of life-changing awe—what I called a zap—and I projected my own feelings of wonderment onto the astronauts. In that sense, I was like much of the public, who had a collective desire for these men to have been somehow transformed by their experiences. I thought

about it every time I looked at the rising moon. "What must *they* think," I wondered, "when they see the moon coming up?" And so, during each new encounter with a moon voyager I pushed—calmly, respectfully—for what I felt certain must be there. But for the most part, I'd met resistance. Still, I couldn't help myself. I was compelled, again and again, to seek what to me had become a Holy Grail.

Dave Scott was the eleventh lunar astronaut I'd sat down with, and he was incredibly generous and articulate in what he shared with me. Sixteen years earlier, in the summer of 1971, he and Jim Irwin had become the first humans to visit the mountains of the moon. For three days they lived in a majestic valley at the base of the towering peaks of the lunar Apennines. With their battery-powered Lunar Roving Vehicle—the first car on the moon—they drove for miles across a spectacular sun-drenched landscape under a pitch-black sky. In search of geologic treasure they ventured to the edge of a winding canyon, to the rims of giant craters, and hun-dreds of feet up the side of a mountain, where, in stolen moments, they'd gazed out on a bright, alien wilderness. During our conver-sations I felt my mind overflowing with wondrous details, ecstatic about the gems on my tapes. Scott even told me, "No one else has dug this deep." But despite the treasure trove of recollections I was amassing, I still felt compelled to dig for something deeper, some-thing *transcendent*. And now, sitting across from Scott, I saw the same familiar reaction—a trace of irritation, even frustration—to my attempt to shape *their* narrative. It was another reminder of what kind of men the Apollo astronauts really were.

———

BORN BETWEEN 1923 AND 1936, THE MEN OF APOLLO were children during the early days of aviation, captivated by the exploits of Charles Lindbergh and the flying aces of World War I. They grew up to become jet pilots, some in combat, others in peacetime squadrons, but always under the looming threat of the cold war. And many ascended to the rarefied atmosphere of test flight, risking their lives to gather data on new, untried aircraft, flying the hottest and most dangerous planes around. All were overachievers who wanted nothing more than to tackle a job almost no one else could accomplish. And so when the space age began and the word "astronaut" entered the language, it seemed as if this fantastic new enterprise had been made for them. They would be fliers of a different kind, beyond the atmosphere. They were just the right age, at just the right moment in history, to become the first humans to leave the Earth and journey into deep space. As military men they welcomed a mission of crucial importance to the nation—the moon race with the Soviet Union—and as pilots they reveled in the chance to carry out the ultimate test flights. For some, there was the added

appeal of becoming explorers, of seeing landscapes no human eyes had ever seen, of standing on a place where no one had ever been. And for the one moon voyager who did not come to NASA as a pilot—geologist-astronaut Jack Schmitt—it was a chance to practice his science on another celestial body. For all of them, going to the moon was the pinnacle of their profession.

But when they returned from the moon, they faced a new mission, one they could never have anticipated and had certainly never trained for. As the first humans to visit another world they now had to face the relentless curiosity of their fellow Earthlings, and that meant a lifetime answering the question: "What was it like?" Or worse, "What did it *feel* like?" Ironically, it was the one question they felt least comfortable answering. For one thing, they were all from a generation in which what you *felt* was much less important than what you *did*. And most of them were not only pilots but also engineers, the ultimate left-brain thinkers. As Apollo 14's Ed Mitchell told me, "I didn't know what feelings were." For anyone in this line of work feelings were not only a distraction but also a danger. Feelings could cloud the mind and impair the process of making decisions in high-speed, life-or-death situations. Feelings could kill you. For men who were wired to be the best at whatever they did, it was the worst sort of bind, because unlike a space mission—where every task was written on checklists and practiced until it was almost second nature—this was a mission with *no* objective criteria for success. And, dauntingly, one with no end point: It would be with them for the rest of their lives.

By the time I sat down with them, more than a decade had passed since the last Apollo mission, and most were reluctant to revisit an experience they had described countless times before. They'd also grown understandably weary of trying to be the "poets" the public craved, feeling inadequate to the task. And even those who didn't mind plumbing their own emotional depths pointed out that recording the thoughts and feelings of going to the moon wasn't a priority at the time. "I wasn't there to remember," Apollo 14's Stu Roosa told me. "I was there to fly."

What they didn't realize was how well they succeeded at the mission they never trained for. What they gave me, with remarkable candor, was *wonderful*. Despite any misgivings, many came through with powerful emotions and observations. A few even said they had been changed, in small or great ways, by the experience. In more than one hundred hours of conversations I came away with an extraordinary archive of lunar experiences, told through remarkably varied perspectives. Ultimately, I realized, I'd found my Holy Grail in that rich and surprising trove of recollections.

———

IN THE SIXTEEN YEARS SINCE MY LAST INTERVIEW, with Apollo 16 command module pilot Ken Mattingly in August 1992, I have thought many times of those incredible conversations with the moon voyagers. In the last few years I've been dazzled by NASA's beautiful new high-resolution scans of the photographs they took during their missions. They are so good that you can zero in on a small section of an image, bringing yourself into the scene as never before. Seeing their explorations in unprecedented detail, I found myself wanting to hear their words, to bring these images fully to life. This was the inspiration for *Voices from the Moon*.

I enlisted my wife, Victoria—a writer and editor whose fascination with Apollo had brought us together—to comb through the interviews and bring a fresh eye to them. Our editorial challenge was complex. Out of a wealth of options we needed to choose quotes that would convey the arc of the entire lunar journey, including the before and after—from the drives that helped steer them on their paths to the astronaut corps to the enduring impact on their lives of their unique status as moon voyagers, and even their views on humanity's future in space. We also wanted to bring each astronaut's distinct personality to the page. And most of all, we wanted to capture the feeling we had, reading the astronauts' words, of being immersed in their other-worldly experiences.

To that end, we became obsessed with combining the quotes

we had so carefully chosen with just the right photographs to magnify the power of the whole. We spent countless hours selecting Hasselblad pictures from among the thousands taken during the Apollo missions, using crops that take advantage of the superb resolution of the new scans. We also screened new high-definition transfers of the astronauts' onboard 16-mm film footage, provided by Mark Gray of Spacecraft Films, and culled scenes not recorded in the still photos. We even did the page layouts ourselves—with the kind indulgence of Viking Studio's design department—so that words and pictures would tell a compelling and coherent story. Remarkably, as we shaped this book, we found that the astronauts' combined voices succeeded as a compelling group narrative that needed no additional words of explanation. On these pages you can hear, as we did, the testimony of this handful of men who have journeyed to the edge of human experience.

Those voices speak revelations. There is Buzz Aldrin's precision in describing the subtle curvature that made him realize he was standing on a sphere. Charlie Duke's sense of wonder at the surprising serenity found in such a hostile place. Bill Anders's awe at the incredible beauty of the Earth afloat in blackness, especially when seen against the moon's lifeless desolate horizon. Alan Bean's amazement at running in the moon's one-sixth gravity and being "airborne" so long that he could relax his leg muscles *during* a stride so that he felt he would never get tired. Ken Mattingly, orbiting the moon alone, over the far side, out of contact with Earth, feeling not loneliness but *exhilaration*. Neil Armstrong recalling the fast-time intensity of humanity's first landing on another world. And Dave Scott, full of detailed description about standing on the side of a lunar mountain and spotting his lunar module, a tiny manmade speck in the ancient, pristine panorama. Each astronaut added essential details to a full portrayal of the moon experience. As we listen to their reflections—sometimes searching for the right words, often remarkably eloquent—their individual personalities and perspectives shine through, each distinctive voice adding unique dimension to this rarest of journeys. I'm only too aware of the silenced voices of the moon voyagers who have passed on: Pete Conrad, Alan Shepard, Stu Roosa, Jim Irwin, and Ron Evans. Their passing makes these recollections even more precious.

—ANDREW CHAIKIN
with VICTORIA KOHL

August 2008

THE APOLLO LUNAR MISSIONS

APOLLO 8 | December 21–27, 1968

COMMANDER: Frank Borman
COMMAND MODULE PILOT: James A. Lovell, Jr.
LUNAR MODULE PILOT: William A. Anders
MISSION: First manned flight around the moon. Borman, Lovell, and Anders made ten orbits of the moon on December 24, 1968.
DURATION: 6 days, 3 hours, 1 minute

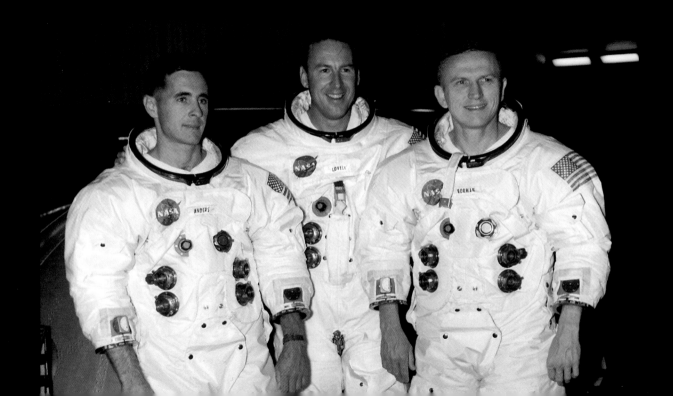

APOLLO 10 | May 18–26, 1969

COMMANDER: Thomas P. Stafford
COMMAND MODULE PILOT: John W. Young
LUNAR MODULE PILOT: Eugene A. Cernan
MISSION: Dress rehearsal for the lunar landing.
While Young piloted the command module,
Stafford and Cernan descended to within nine
miles of the moon's surface in their lunar module.
DURATION: 8 days, 0 hours, 3 minutes

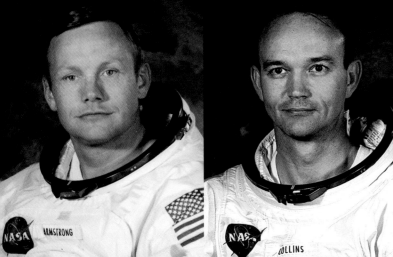

APOLLO 11 | July 16–24, 1969

COMMANDER: Neil A. Armstrong
COMMAND MODULE PILOT: Michael Collins
LUNAR MODULE PILOT:
Edwin E. "Buzz" Aldrin, Jr.
MISSION: First lunar landing. Armstrong
and Aldrin's single moonwalk lasted about
two and a half hours.
LANDING SITE: Sea of Tranquillity
DURATION: 8 days, 3 hours, 18 minutes

APOLLO 12 | November 14–24, 1969

COMMANDER: Charles "Pete" Conrad, Jr.
COMMAND MODULE PILOT: Richard F. Gordon, Jr.
LUNAR MODULE PILOT: Alan L. Bean
MISSION: First landing on the moon at a prechosen spot.
Conrad and Bean made two moonwalks totaling nearly
eight hours, during which they deployed the first Apollo
Lunar Surface Experiments Package (ALSEP) and visited
the unmanned Surveyor 3 probe.
LANDING SITE: Ocean of Storms
DURATION: 10 days, 4 hours, 36 minutes

APOLLO 13 | April 11–17, 1970

COMMANDER: James A. Lovell, Jr.
COMMAND MODULE PILOT: John L. Swigert, Jr.
LUNAR MODULE PILOT: Fred W. Haise, Jr.
MISSION: Third lunar landing attempt; aborted on the way
to the moon after the explosion of an oxygen tank inside the
service module.
DURATION: 5 days, 22 hours, 54 minutes

APOLLO 14 | January 31–February 9, 1971

COMMANDER: Alan B. Shepard, Jr.
COMMAND MODULE PILOT: Stuart A. Roosa
LUNAR MODULE PILOT: Edgar D. Mitchell
MISSION: Third lunar landing, and first successful mission devoted entirely to scientific exploration of the moon. Shepard and Mitchell took two moonwalks totaling more than nine hours.
LANDING SITE: Fra Mauro highlands
DURATION: 9 days, 0 hours, 2 minutes

APOLLO 15 | July 26–August 7, 1971

COMMANDER: David R. Scott
COMMAND MODULE PILOT: Alfred M. Worden
LUNAR MODULE PILOT: James B. Irwin
MISSION: Fourth lunar landing and first visit to lunar mountains. First extended lunar scientific expedition featuring extended lunar stay time, long-duration backpacks, and battery-powered Lunar Rover. Scott and Irwin took three moonwalks totaling more than eighteen hours. The Apollo 15 service module was the first to be equipped with the new Scientific Instrument Module. Worden performed a thirty-eight-minute spacewalk to retrieve scientific film during the trip back to Earth.
LANDING SITE: Hadley-Apennine region
DURATION: 12 days, 7 hours, 12 minutes

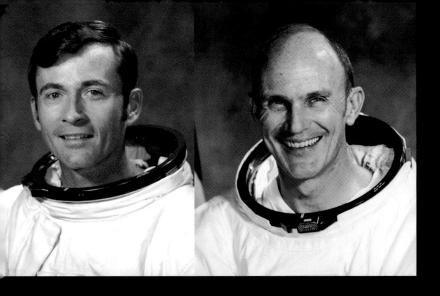

APOLLO 16 | April 16–27, 1972

COMMANDER: John W. Young
COMMAND MODULE PILOT: T. Kenneth Mattingly II
LUNAR MODULE PILOT: Charles M. Duke, Jr.
MISSION: Fifth lunar landing and first exploration of the moon's central highlands. Young's and Duke's three moonwalks totaled more than twenty hours. On the return trip Mattingly took a spacewalk lasting 1 hour, 24 minutes.
LANDING SITE: Descartes highlands
DURATION: 11 days, 1 hour, 51 minutes

APOLLO 17 | December 7–19, 1972

COMMANDER: Eugene A. Cernan
COMMAND MODULE PILOT: Ronald E. Evans
LUNAR MODULE PILOT: Harrison H. "Jack" Schmitt
MISSION: Sixth and final Apollo lunar landing. Geologist-astronaut Schmitt became the first professional scientist to land on the moon. First U.S. manned night launch. Longest Apollo flight. Cernan and Schmitt took three moonwalks totaling twenty-two hours. The spacewalk by Evans on the return voyage lasted 1 hour, 7 minutes.
LANDING SITE: Taurus-Littrow valley
DURATION: 12 days, 13 hours, 51 minutes

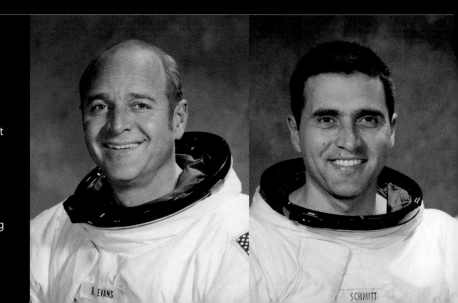

VOICES FROM THE MOON

1

BEFORE

I would've been reluctant to accept in the middle 1950s that we would see
spaceflight in my lifetime.

—NEIL ARMSTRONG

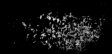

I was very much interested in the stars, the moon, all that kind of stuff. That held my interest. . . . Why was I so interested in it when I was young? It was just *fascinating* to me, absolutely fascinating. Maybe it's because it's sort of unknown. Maybe because every night you can see the stars, and you fantasize or dream, or something like that, hear people are going to do that. I mean, it probably holds so much more interest than even, you know, kids growing up being airplane pilots, or railroad guys, or something like that. It's such a fascinating science. Such an interesting thing that captures the imagination of more and more people than anything I can think of. . . .

Before there had been a NASA, or thoughts about going to the moon, before there had really been missiles, I was very much interested in rocket power. . . . Just before I graduated from high school, my first love was to be a rocket engineer. And I had written to the secretary of the American Rocket Society requesting to find out how to become a rocket engineer. And he had written back that there wasn't such a discipline at that time, but if I take mechanics, and thermodynamics, and math, and things like that, they would prepare me in the future. And he also suggested that I go to Caltech or MIT, but I couldn't afford either one of those two. So I actually eventually ended up at the Naval Academy, where, in my first class year, I wrote [about] the development of the liquid-fueled rocket engine . . . [for] the term paper or thesis that everybody had to write. You could choose any subject, but that was the subject I picked. So essentially, what I'm saying is that in 1951 I had been very much interested in what eventually I got into in 1962. Long before the Glenns and the Shepards and everybody else had even thought about the space program.

—JIM LOVELL

Above: *A space-struck Jim Lovell as a Boy Scout.* Right: *the future crew of Apollo 8, from left, West Point cadet Frank Borman and Annapolis cadets Jim Lovell and Bill Anders.* Opposite: *Stu Roosa during air force flight training.*

remember my first ride in an airplane. I was about five years old. . . . It was a Waco cabin plane. And I was in Ambler, Pennsylvania, at an air show. And I just badgered the shit out of my old man to pay five bucks or whatever it was in those days to go for a ride. And I beat all the old folks and everybody else into the right front seat of that thing. So I was up there where the controls were and everything, and, Christ, I couldn't have been big enough to either see out or reach over them or anything. But that was my first airplane ride, and that was the most natural place in the world to be.

[As a kid] I was thinking about guys who flew fighters in World War I; at that time World War II hadn't come along. And I thought about the Lindberghs. I used to build paper airplane cockpits on the floor and around chairs and sit in 'em and pretend I was flying, for hours—not space. But it's the flying that got me to the right place at the right time. I just happened to be the right age, let's face it.

—PETE CONRAD

I enjoy flying. That's all I wanted to do. When I was a little kid, I wanted to fly. Wanted to build airplanes and fly 'em. So, [at NASA] they gave me some funny airplanes that didn't have wings, but it's still the same thing. . . .

—KEN MATTINGLY

If they made a movie about your book, . . . I would probably be the classic one . . . redheaded, freckle-faced kid sitting in Oklahoma, looking up at the moon and talking with his dad that I'd like to go to the moon. Well, that's the Hollywood version. That's not true with me. I have a close parallel to that: A redheaded, freckle-faced kid sat on a small porch in Oklahoma and talked with his dad about flying airplanes. And I have wanted to fly airplanes *ever since* I can remember. I cannot remember back when I suddenly decided I wanted to fly airplanes.

—STU ROOSA

You come back from a [navy] test hop . . . and think, Boy, I'm really hot shit! I did this today, and it's really good. . . . So you feel like, Hey, I'm really doing good stuff for the navy, and, Probably nobody else in the world could do this. . . . That's the feeling—that's what made it so great.

You know, when you're saying, Should I risk all this? And I still have a young son and daughter and wife. Then you're saying, Well, you know it would be awful selfish if I did all this because I want to fly higher and faster than anybody. . . . You know, it's pretty selfish if you think about it just in the abstract here, without anything else. But if you look on to the end—We're doing this for the good of our country—then you're not being so selfish. In fact, you might even consider you're being self*less*, because you're risking your life in a worthy cause.

—ALAN BEAN

It came down to two words: *higher* and *faster*.

—TOM STAFFORD

The kind of personality that enjoys this—and I don't single it out as being anything laudatory—it's just there's a personality set that enjoys this business, that says the most exciting thing in the world is to do something that somebody hasn't done before. And it's based on our best calculations and analysis, imperfect as they are; it's the best we can do. And we think we have enough confidence in it, that we'll go try it. And "try it" means, now that I understand it, am I willing to hang my neck on it? Now, I find making that kind of decision . . . the epitome of opportunity. . . . It isn't academic, it's not a quiz. It's—Okay, it's your life. You gonna ride it? You quickly know what your answer is, and you quickly know where you are. And I thought, You know, that's about as neat a life as you could ever have.

—KEN MATTINGLY

Clockwise from left: *Tom Stafford as an instructor at the Air Force Test Pilot School; navy squadron pilot Alan Bean; air force test pilot Mike Collins.*

The *mission* of our nation was to send a man to the moon and return him safely to the Earth. . . . And I was part of that mission. So I was in a position where I could, instead of shooting, being part of a war—and by the way, I came out of Vietnam—I could really do something for my country, and I didn't have to be fighting a war. . . . I flew a hundred missions in Vietnam. . . . Let's put it this way. Anytime you're flying and people are shooting at you, it sort of changes your outlook about the enjoyment of flying airplanes!

—RON EVANS

I went to test pilot school because it was the thing to do. . . . Man, I'd been on the gunnery team as a fighter pilot, and I flew on the wing of the squadron commander. I was hot stuff: I mean, I flew well. I was lucky, I had a good instructor when I went through flying school, and I won all the prizes. . . . So I figured, shoot, I'll go to test pilot school and get that on my résumé. . . . That's where I *really* learned to fly. Because that's absolute precision flying, and they have you do things that you just don't normally do in the squadron, and you had to do. So it was a big difference between a test pilot and a pilot, in terms of handling flying machines.

And you gotta remember, back in the early days, in Mercury, and Gemini, and early Apollo, we didn't know what kind of flying machines we had. Nobody knew how to fly them. That's why we were there—to help design how you fly a space ship. What do you put in a space ship? How do you make it work? How do you make it go this way and that way? What do you put on the panel? That's why they had engineering test pilots, and that was exactly the right decision.

—DAVE SCOTT

Right, above: *Ron Evans, aboard ship in Vietnam, learns he has been selected as an astronaut. Right: Dave Scott as a test pilot at Edwards Air Force Base. Opposite: Frank Borman at Edwards.*

I had one goal—that was participating in the program and making sure that the program was a success. I felt, honestly, [Apollo] was an enormous national drive, an enormous national endeavor. Like a war. I wanted to see the war won, and I wanted to participate in it, but I wasn't trying to be the guy that carried the final ball. . . . As far as whether I had a personal goal to walk on the moon, the answer's no.

Again, it was a mission to me. If I hadn't been there, I would have been somewhere else doing it. I mean, I would have been in Vietnam. I was, at that point in my life, very, very oriented toward doing what was expected of me and doing it better than anybody else can do it. I was a military officer, and . . . if they had sent me to Vietnam, I would have tried to be the very, very best fighter-bomber pilot, or whatever they told me to do. . . .

Landing on the moon, to a lot of us, wasn't the be-all and end-all of life. It may have been to some of 'em, but I never had that romantic [attitude]—I had a goddamned job. You could have just as well told me . . . to bomb a bridge or drop an atom bomb, and I would've done it with the same enthusiasm. Okay?

—FRANK BORMAN

I don't think I'm that much of a philosopher. I don't think I have any big philosophical, religious, or any other motivations. . . . I mean, the program is, we're going to send people to the moon, and I want to go.

—STU ROOSA

I didn't have some great worldly philosophical bullshit about going to the moon. It was just, you know, Goddamn, this is gonna be neat!

—PETE CONRAD

2

PREPARING

That's always in the back of your head—you never really think you're quite ready. . . . You just train the best you can, pile it in. And one day you say, We're gonna go. . . . I'm as ready as I'll ever be.

—AL WORDEN

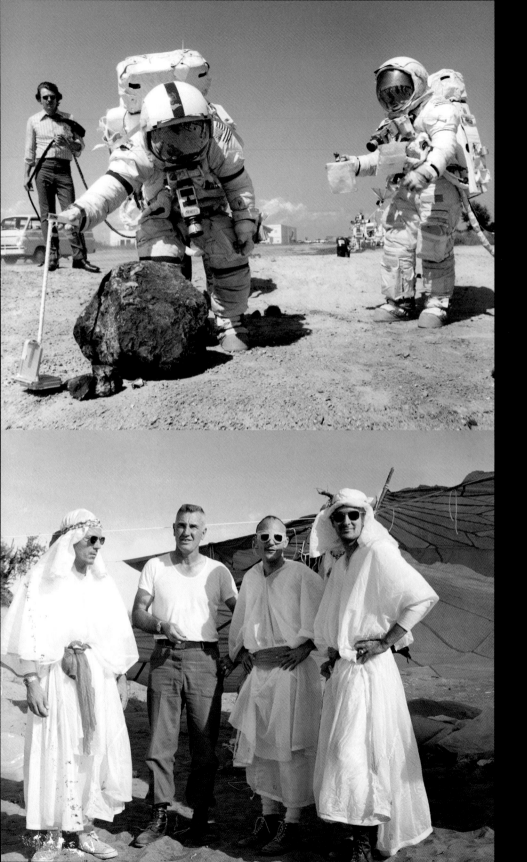

It was [the simulation instructors'] job to make very difficult situations, and they were successful in doing that. But the result of that was that the flights were always easy because they were so much more *(laughs)*—more peaceful than the simulations that you considered it a welcome change.

—NEIL ARMSTRONG

I think it's like any team. You take a couple of quarterbacks in the National Football League that have been playing together for four years. Do they anticipate the other guy's moves? Absolutely. Do they know where the other guy is every minute? Absolutely. That's what makes a good defensive backfield. 'Cause they all know where the other guy is. So, I knew Jim's moves, and he knew my moves. I mean, that's why you do all that together. I mean, gosh, we flew together, we trained together, we went everywhere together. . . . We were totally together for a long, long time.

—DAVE SCOTT

The Rockpile was pretty exhausting . . . a combination of the Florida heat, the weight of the backpack. . . . On the moon, we didn't have the weight, [but] we had the pressure of this being the real thing and not a simulation, and then we had a stiff suit to contend with. We were always fighting the suit.

—JIM IRWIN

It's always hard work running those inflated suits around. And in one G it's worse. Nevertheless, I thought those were always worthwhile exercises. Appreciated the chance to do them.

—NEIL ARMSTRONG

Clockwise from left: *Jack Swigert, Ken Mattingly, and Charlie Duke during survival training with an instructor; John Young* (red stripes) *and Duke practice a moonwalk on the "rockpile" at the Kennedy Space Center; Ed Mitchell and Al Shepard train for one-sixth gravity on a KC-135 aircraft; Pete Conrad and Alan Bean in the lunar module simulator; Mike Collins ready for a centrifuge run.*

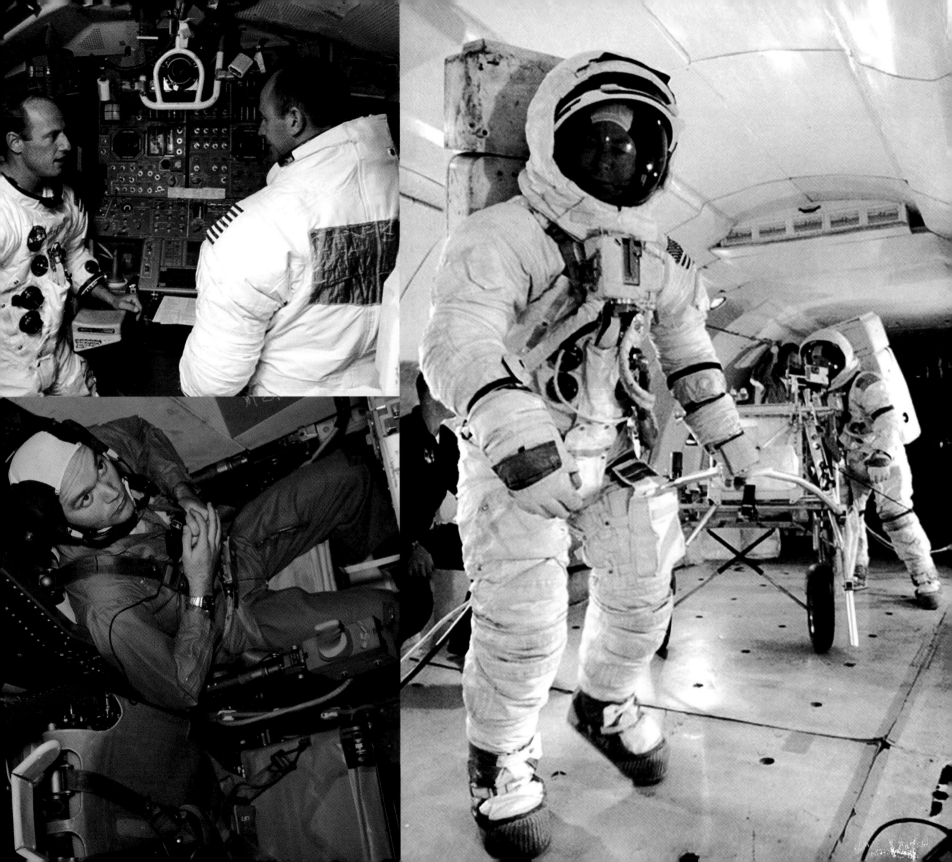

It was a fairly unforgiving machine, it's true. For that very reason, it had merit *(laughs)*. Because—I mean, it was the kind of pressure that you would experience in the lunar module. . . . I don't think it became second nature, but it became fairly comfortable as a result of the [Lunar Landing Training Vehicle] activities. We felt confident that we could fly to a specific spot and land there, and that was important. . . .

From my point of view, the tough parts were the unknowns, and the concerns about things that might come up that you didn't properly anticipate. You know, you fly an aircraft, and you land on a runway, and you know about where it is, and what it looks like when you're close, and what the touchdown feels like, and when you're in some kind of trouble. And you didn't have those senses about the lunar module. So in that sense it was tougher, because you had less experience to draw from. And that's why I felt it was important to have as much realistic experience as we could before going there, and that's why the LLTV was important, in my mind.

—NEIL ARMSTRONG

When you fly all your life in air, then air becomes a part of the equation. But in this situation, it's not part of the equation. So you have to learn those motions.

—DAVE SCOTT

Astronauts who flew the demanding and dangerous Lunar Landing Training Vehicle, or LLTV, included Dave Scott (left) and Pete Conrad (opposite).

[The simulator] doesn't do the physical phenomena, and it doesn't put you on the line. The LLTV puts you on the line. Because you know that at five hundred feet, if you screw up, you're dead. And *that's good*. . . . It makes you focus. In the simulator you can say, We gotta run an altitude chamber test this afternoon. Or, I gotta go check my suit. Oh yeah—I'm landing now. You could let your mind drift. It's no big deal. Flying the LLTV, you don't let your mind drift. So it's good training, 'cause it puts you on the line.

That thing was a hell of a challenge to fly. Because it's very demanding, and things happen very quickly. . . . If you want to stop a helicopter, you pull the nose up, and it will stop, and it's floating on the air. In the LLTV, when you pull the nose up, it doesn't float in the air; it falls down. . . . It was very unforgiving, because if you screw up it crashes. . . . But the beauty of it is, it teaches you to run your [mental] computer fast.

—DAVE SCOTT

I think the LLTV was just a confidence booster. Really *(laughs)*. You know, you get the thing off and back down again, and you say, Well, walk away from that one. . . . Some airplanes you can fly and make some mistakes and get away from it. That thing, I mean, you really had to be on top of it all the time, or it was the end of the flight. So toughness, in the sense that it was tough to fly, but also mentally tough, to be that close to the edge all the time. You know what I mean?

—ALAN SHEPARD

I liked the fellowship. I really liked the people. I remember we went out in the Oricopias [mountains in southern California] for six days. . . . Did some good traverses. Had some good times. Sit around and shoot the breeze in the evening. To me that's very enjoyable. I like that. . . . The things you like you do well. You go back and you look at your academic background—where did you learn the most? From the teachers you liked. . . . So you go out in the field, you have teachers you like, you learn a lot. . . . I looked forward to having those guys [geology instructors] come in and tell us things. Here's where you're going, here's what it might be, and this is why. Great mystery! Great story, huh? . . . We're going to go solve a mystery!

I felt like we needed those guys to help us do what we're supposed to do, and it becomes a team effort. I never went on a field trip without a [Bill] Phinney or a [Gordon] Swann or a [Jim] Head. . . . Those guys were always, always there. So you get comfortable and you want them to be there. Even if just by radio . . . 'cause they're sort of your buddies, the guys who can help you out. You know? This purple rock—"Hey, Lee [Silver], tell me what to say!" . . .

It just sort of grew on us. Because no trip was like the previous trip. They were always different. That was good. . . . We learned to learn. We learned variety. We learned to take the unexpected. . . . There was constant emphasis and focus on what we were doing. That's the way [Lee] Silver operates. . . . There wasn't a sit-around-the-fireside-and-talk-about-airplanes. We talked about geology, talked about rocks—this rock or that rock, this site or that site, this morphology. There was a constant focus on the geology, and I think that was very good. Because he made it interesting . . . and we were becoming prepared for what we had to do.

—DAVE SCOTT

Clockwise from right: *Geology training at New Mexico's Taos Gorge for Dave Scott at the canyon rim; Scott and Jim Irwin in a training version of the lunar rover; Apollo 16 prime and backup crews with geologist Lee Silver (pointing); geologist Farouk El-Baz (glasses) trains Ron Evans while support astronaut Bob Overmeyer listens.*

You're remotely involved in it, and then it gradually grows, and before you realize it, it is gonna happen. So that, in that gradual way, leads one into accepting as commonplace, as normal, as routine, what to someone else is phenomenal, what is fantastic, what is—What an opportunity! Once in a lifetime! Ah, shit, it's routine. So what else is new? We're going over and training for landing on the moon. "You're what?" Well, we're gonna train on landing on the moon. But that becomes routine after awhile. And you want it to be that way. You don't want it to be a big deal that clouds your ability to handle it as a normal [thing]—'cause the best way to deal with a difficult task is to treat it as a normal event.

—BUZZ ALDRIN

That's part of the business of being a pilot, is not to get excited particularly about things, so that you can function. It's the Chuck Yeager stay-cool routine, where you . . . everything's okay. Things can be going to hell around you, but you stay cool. Because in theory, you've got to be cool to make the right decisions. So I think awe prevents you from making the right decisions; fear does, all these other things. You're very susceptible to all these things, but you've got to not be. And so I think one of the things that you have learned through the years is to stay cool. So when you see something amazing coming up, you kind of put a damper on it, so you can stay cool and stay in control. . . . The ones that are the good guys are the real cool guys, you know—their plane was on fire, and they did this and this. They talk real calm. . . . See, even in World War II movies it was true. The really good guys, like John Wayne, he doesn't say much. John Wayne doesn't ever panic.

—ALAN BEAN

The entire environment in which we were working was a fishbowl, and everybody was watching, and the press was extreme. It would've been impossible for us not to be aware that (laughs) everybody was looking over our shoulder.

I was getting more aware of spiritual truth before the flight. I wanted to be prepared in every way in case I didn't come back. I knew there was that possibility. I wanted to know, first of all, where was I going when life ended. If you're ready to face death, then you're more ready to face life. You do things without worrying, because you've already put that behind you; you've faced up to that.

—JIM IRWIN

Somewhere along the line I made a rational, coldhearted decision that, on the one hand, there was a one-in-three chance of not making it back. And I wasn't just worried about me; I thought about my family. We had no insurance; you couldn't get insurance. And we had five kids, and how irresponsible it would be. . . . It certainly was a negative on that side of the balance. Not to mention my own hide. Versus the other things, which were . . . adventure, exploration, national prestige, personal honor, and excitement. . . . And I had decided that that was acceptable.

—BILL ANDERS

We'd look at [the moon] at night, and talk about it. But it still seemed pretty abstract. It's funny how you get. When you're planning on going to the moon every day, it seems . . . a little bit in the future and unreal, because you've been saying you're going to the moon for weeks and weeks and months and months. And you flew to the moon in the simulator yesterday and the day before. In fact, when you go down to the launch pad, until about fifteen minutes prior to launch . . . you know, you're saying, Well, maybe we'll go, maybe we won't. Here you are, it's launch morning, you've got your suit on, you're in there, but you know, you've done this so much that . . . it's not unlike all these other practice times. And it's only about—for me, anyhow—about fifteen minutes prior to the real launch that . . . I began to really change from doing the job to We're really going.

—ALAN BEAN

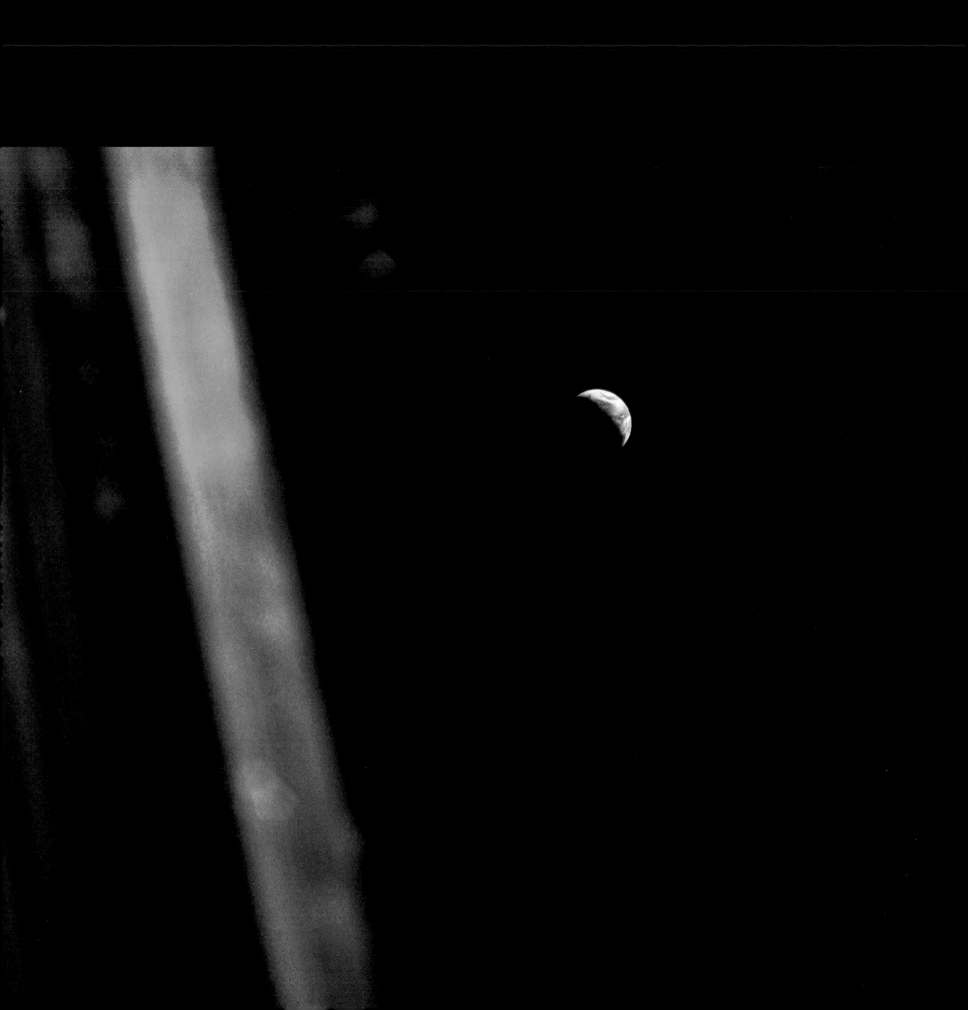

3

OUTWARD BOUND

You say, Hey, I'm out here 150 or 200 thousand miles away from home, going in the other direction. It's not just home—it's not like you're on a trip from Houston to California. I mean, you have really left *society.*

The Saturn V was such an enormous machine. And the size of the engines?! You still wonder, when you see it on its side down there in Houston. It was an enormous thing. And I think I felt that more going up the morning of the launch. Because it was so quiet, nobody around it. . . . I don't want to say awe, a combination of admiration—yeah, maybe awe. Wonderment.

—FRANK BORMAN

It's a little different sitting in the rocket, rather than watching it . . . from the ground, and hearing the announcer, you know, dramatically talk about the countdown, and what's going on. Inside the rocket, sitting there, waiting for the countdown, is a lot different, because you don't get that momentous buildup, that anxiety buildup. You're sitting there, and you just do certain things. And the launch is a little bit different too, because on the ground you get that vibration in your stomach, whereas in the spacecraft itself, it's a big rumble. You can hear those valves open up and all that fuel drop down those manifold valves. You know, the pipes are big. You know, you're burning fifteen tons per second. And so you really go to town, and you can hear that. And it's a big rumbling noise, and off you go.

—JIM LOVELL

There's always the element of unreality in it because a launch is not real until you lift off. And until you lift off, something could always happen to call you back, to prevent the launch. . . . So you don't commit yourself to the flight—totally—until you get ignition and you're off the pad. And *then*, it's all or nothing. That's the gamble—it's either heads or tails. At that point, you're committed to the flight. Whether you come back is not important at that point. Then, the flight is the important thing. . . . I would say, at the instant of liftoff *(snaps his fingers)*, and you know they can't call you back, there's a momentary thing that says, This is for real. And then, training kind of takes over. And you go through things like you did in the simulator.

—AL WORDEN

Above: *Apollo 8's liftoff, viewed by a camera on the launch pad. Opposite: A tracking camera view of Apollo 15.*

There was a startling moment there, right at liftoff. Everybody got quite startled. Because we had simulated the hell out of everything—aborts and everything—but nobody had ever been on a Saturn V. . . . As we lifted off, you can imagine this rocket—it's a giant thing, but it's not bulky like an obelisk or like the Washington Monument; it's not rigid. It's more flexible. Not quite a whip antenna on your automobile, but somewhat like this. . . . So we were literally being thrown around. I mean, "thrown around" is the best way I can describe it. I felt like a rat in the jaws of a giant terrier. I mean, here we'd hardly started, and already we had something that we hadn't simulated.

—BILL ANDERS

I really wasn't sure the crazy thing was going to stay together. Even to read the gauges was almost a guess.

—RON EVANS

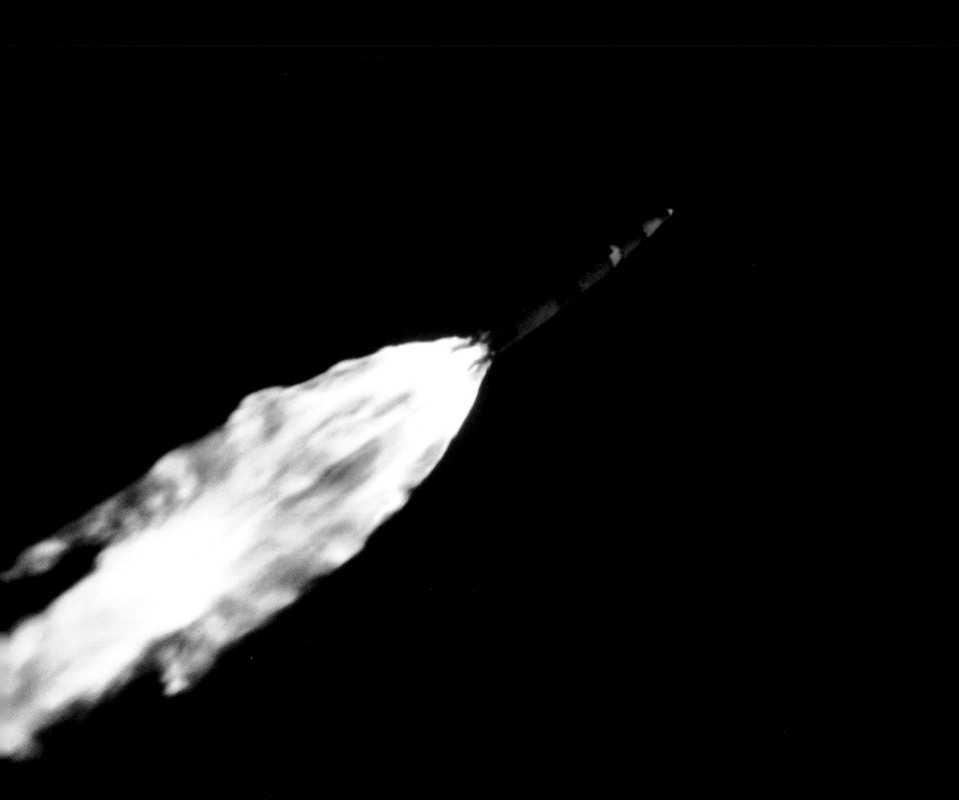

It was raining so goddamn hard—it was really a damn storm that morning. We wanted to launch, obviously. We delayed during the countdown, but we weren't about to crawl out of that goddamn thing and go back. We were ready to launch. And then we were running out of the [launch] window, and it looked like it was easing off some, and they fired our butts right through that stuff.

— DICK GORDON

No matter what single, double, or triple failure those guys [the simulation instructors] put into the electrical system, they never came up with anything that turned on every electrical warning light in the caution and warning system. Man, they *all* lit. I think there were eleven of them. And they all came on. Everything that had anything to do with the electrical system lit up on the caution and warning panel. Every one of those hummers was on. *Every one!* I couldn't believe it.

— PETE CONRAD

Pete called it right; he told [the ground] he thought we got struck by lightning, but neither Al nor I had a window to look out of, and we didn't see anything. . . . There was a boost protective cover over us; during launch, his is the only window . . . until the [launch escape] tower goes and pulls the boost protective cover off.

— DICK GORDON

I thought the service module had somehow separated from the command module. Because I didn't know any other way—I knew that no failure, or two failures, could do it. Because we'd had all the failures. So I knew them. You know, I'd look at six lights; that's AC [bus] 1. You soon learn the patterns and the numbers. And there were so many. . . . I said, "They didn't bolt the command module right to the service module, and it slipped." Because, see, we lost three fuel cells. The only way you can do that is to kind of break it. . . . So that's what went through my mind. I never thought of a lightning bolt. . . .

— ALAN BEAN

I never thought about aborting—at that point. Obviously, I did not want to wind up with a dead spacecraft in orbit.

— PETE CONRAD

In retrospect, it could have been catastrophic. But it wasn't.

— DICK GORDON

We had a lot of acceleration just prior to [first-stage] cutoff. We were really being squashed back. . . . We were up to four and a half Gs or whatever it was. And, you know, your chest gets compressed down. You're panting. Your arms feel real heavy, so you're not moving around flipping any switches. And of course the fluid is all back here in your ears. But you get used to it. So you're kind of semiacclimated. And suddenly, you go from that, not only to zero G as the engine cuts off, but there's little retrorockets that fire on that engine to pull it back off, just before the second stage cuts in. . . .

You know, you've seen those old movies like *Captain from Castile*, where they have a catapult that heaves the rock over the wall? I mean, I suddenly felt like I'd been sitting on a catapult and somebody cut the rope. Because I felt like I was going to go right through the instrument panel. Literally. . . . And so I threw my arms up. And just as I got my hand up like that, the second stage cut in, and, *clunk*, the wrist ring hit my helmet. So I was a little embarrassed. Of course there was this big cloud of fire around us, you know *(laughs)*, it was a very spectacular part of the flight. And of course, I'd just gone through my first launch; then two minutes and forty seconds later, we're in the middle of this, and I thought, Boy, this is going to be something. [It was] dull after that.

—BILL ANDERS

Having that whole mission in my hands when we lifted off—I had that T-handle, which could've shut that Saturn V down, aborted the mission if I wanted to. I mean, I had that decision to make— anytime, I could've made it, good or bad. You almost wish you had a guidance failure at liftoff. Because I knew I could've flown that big Saturn V into orbit goddamn near as good as the computer.

—GENE CERNAN

Opposite: Apollo 12 lifts off into a rainstorm. Half a minute later, the ascending spacecraft was struck by lightning, knocking out the command module's electrical system.

Right: Burning in the invisible flame of the Saturn V's second-stage engines, a connecting ring falls away following first-stage separation. An automatic camera aboard the unmanned Apollo 4 captured these views.

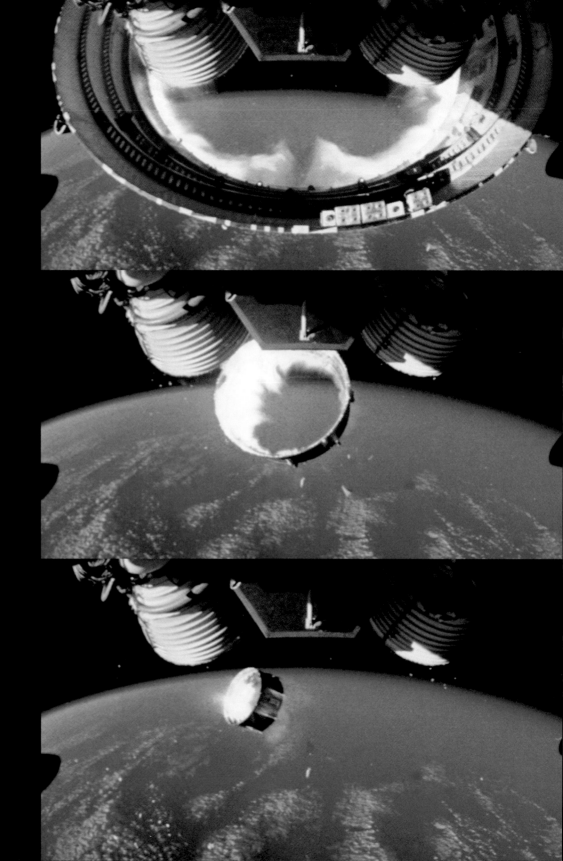

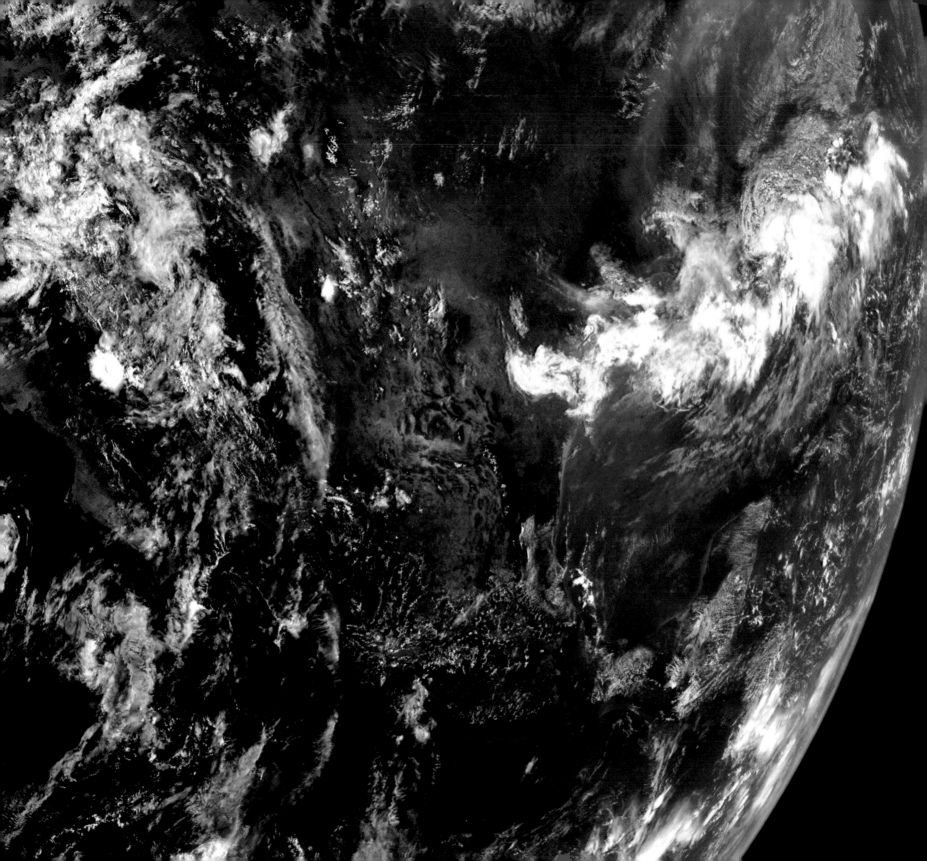

You know, in Earth orbit the horizon is barely curved. All of a sudden you move out at 25,000 miles per hour, and the first few hours, things *really happen*. . . . I mean, you can *see* yourself leave the Earth at a *tremendous* rate of speed. You can see the horizon begin to close in upon itself. You can begin to see the continents. You begin to see things from the top down. You begin to see and realize after a period of time that the Earth's rotating, because the continents are beginning to change places. And the second day, now you've been looking at the Earth, it's become quite small and continues to get smaller, but very slowly does it continue to get smaller. So it's pretty dynamic in those first twelve hours—that's when things really happen.

—GENE CERNAN

In spaceflight, when we orbited the Earth, we thought in terms of continents. We were over the U.S.; now we're over a body of water. We're over Africa now; we're over Australia now. In the lunar flight, we thought in terms of bodies. The moon's here, the sun's there, the Earth is there.

—JIM LOVELL

Mexico dominates this view from the departing Apollo 11.

I fancied myself as a guy who understood geography. And I looked out there, I could *not* figure out what was up. . . . I mean, everybody knows that north is up, right? You sit in the classroom in fourth grade, and you look up there, and the teacher has a globe. There were several things that came across later, and I thought, Jeez, I should have known that. One, the Earth is not divided up neatly into little colored countries. Okay? So you don't see a red America, and a green Chile, and a purple China *(laughs)*. . . . I expected more visual clues as to what I was looking at. Secondly, it's covered with clouds, so that obscures things. And God does not necessarily say that when you look at it the first time that north is going to be up. And it took me like several minutes to finally realize that what was up, was really *Antarctica.*

And I thought to myself, Now wait a minute. Let's go back. What do you see? Well, you see a big white patch. Is that clouds? No, it looks like ice. A big ice patch in the wintertime that you can see, it's got to be Antarctica. Antarctica up? Oh yeah, that could be, cause *we're down.* Well, then, we must be looking at it like *this.* So I actually went and looked at it like that [upside down]–*Yeah,* that's right! That's Antarctica! And then I said, Well, if that's Antarctica, let's start working from there. That thing here. . . . What could that be? My first thought was, That must be the horn of Africa. See, here's the horn, here's Cape Town. Well, if that's Cape Town, where's South America? And what is *this* thing? Then I got to realizing, That ain't the horn of Africa; that's the coast of Chile! Isthmus of Panama, here's Florida. And *here's* Africa. Then it jelled.

— BILL ANDERS

The Earth is [fifty] times brighter than the moon, because of the reflection of the sun's rays on the clouds. But you don't get that on photographs.

— JIM LOVELL

The other thing . . . was that this little spot, the Bahamas lowland, was a turquoise *jewel* that you could see all the way to the moon. . . . It was like it was *illuminated,* like a piece of opal. And you could see that all the way. And I kept being *amazed* about that.

— BILL ANDERS

To me, it's crystalline. Crystalline being it has depth. I like to draw the analogy with someone who has deep blue eyes. . . . The Earth is deep blue. And especially when you get out a little ways, not too far away, and you can look back at it, it's deep blue. It's got a three-dimensional *feel* to it. A depth. And it's really beautiful. . . .

— DAVE SCOTT

I was just wishing I could spin it around and look at the rest of it.

— BILL ANDERS

An Apollo 12 view of Earth includes the Bahamas (turquoise spot at right of center).

The receding Earth, seen from Apollo 15.

You can see the whole Earth at about ten thousand miles. And you start taking pictures. You take one at ten, and one at fifteen, and one at twenty, et cetera, et cetera. And of course, they're all the same; it's just that the Earth takes less of the field of view of the camera as you get further away. But you don't think that. You think, Oh, I wanna take another picture now. I wanna take another picture

It was kind of like, Yeah. The Earth's getting smaller. In fact, this was something that really surprised me. Here you are, watching the Earth shrink. And you know when it really dawned on me that we're a long way from home is when you start picking up the delay in the communications. Now, why looking out of the window seeing the Earth shrink doesn't do it, but why the audio of the delay in the communications does, [I don't know], but it did. . . . When you would call, "Hello, Houston," and then there would be, *Mmmmmmmm* "Go ahead, 14." And that was the first big realization that, hey, we're starting to get out here. More so than seeing the Earth shrink. And I got out of my suit first, and I was flipping around, thinking, Isn't this fun! And then suddenly I thought, My God, if I do this about three more times, I'm going to embarrass myself. So I'm going to quit doing it. . . . I didn't throw up, but I thought, I'd better be careful or I'm going to throw up. . . . After about eight hours, I'd adapted. Reasonably.

—BILL ANDERS

I was really, really worried about [whether I'd get sick in zero G]. And I remember the exhilaration the first time I released the lap belt, got out of the couch, and I thought, Oh God, now we'll find out. And it took about ten nanoseconds to recognize, I've been here all my life. This is *absolutely natural*. And I never gave another thought to it. . . . I must have beamed from ear to ear when I realized, Got it made. This is *perfect*. I know exactly where everything is. Upside down, right side up, it looks perfect to me. It's beautiful. I can move anywhere I want, I can do anything I want, and there aren't any problems associated with this business. And I remember what a euphoric feeling that was.

—KEN MATTINGLY

We lit the [Service Propulsion System] engine to take us off the free-return trajectory. So that's the first time that you light the SPS engine. And item number one on the SPS burn checklist is "Secure all lose items." Okay, so now, you've just spent a thousand hours in the simulator, and you've gone through this how many hundreds of times? . . . Okay, items secure. And then you go on down. And then you're into the nitty-gritty, you know, you get your fuel cells up, and your gimbal motors on, and this check and that check, and you're ready to burn. . . . It was a short burn, I think two or three seconds, or whatever. . . . And so as soon as the engine lit, I was really surprised. Because it—*Pow!* And man, you went back [in your seat], and a checklist goes flying over your head here, and something else goes flying over there. After that, you paid more attention to item one, "Secure all loose items."

Zero G is a blessing and a curse. I mean, for keeping track of your film, it's a curse, because the goddamn stuff, you put it down, which is stupid to do. I always used to put it on the edge of the simulator, and it just stayed there, you know? *(laughs)* And without thinking, I didn't stick it on the Velcro. Put it down here—where is it? I'd have to go hunting for it, and that always puts you a few minutes behind. It's also a curse from the bodily functions. I mean, next time you go to the latrine, imagine if you were in zero G. What does that stuff *do*? It just—*ugggh*. It's hard, even with KC-135 [zero-G aircraft] flights, you just are not able to totally train yourself for zero G until you get there.

The plus side is, it's a very comfortable place. I mean, I can't sit still; my back starts hurting. And yet, I never got uncomfortable on the flight. . . . It's very relaxing. It's *easy*. You don't have to be strong. . . . Mainly you're just like a big fish, like a jellyfish, lying there. And your arms are like this, in their natural position; your legs are like that. . . . Pretty soon, after three days, you're adapted to it. . . . I enjoyed zero G.

The thing that was the most difficult for me to sleep was that the damn sleeping bag was sized for somebody like C. C. Williams, the biggest guy around. I was like one pea in this pod. I didn't realize it, but I kind of like to feel the security of the bed up against me. And I can't sleep on my back; I've got to sleep on my stomach, and feel pressure. And there was no pressure. And I'm sure you've had this feeling, where you're lying in bed, and just before you drop off to sleep, you suddenly feel like you're falling. . . .

—BILL ANDERS

No matter what I say, anybody says, about weightlessness, you cannot relay that to somebody who hasn't had the experience. You just cannot relate to it until you've had the experience of being in zero G. It's absolutely delightful. But how are you gonna explain it? You can't explain it!

—RON EVANS

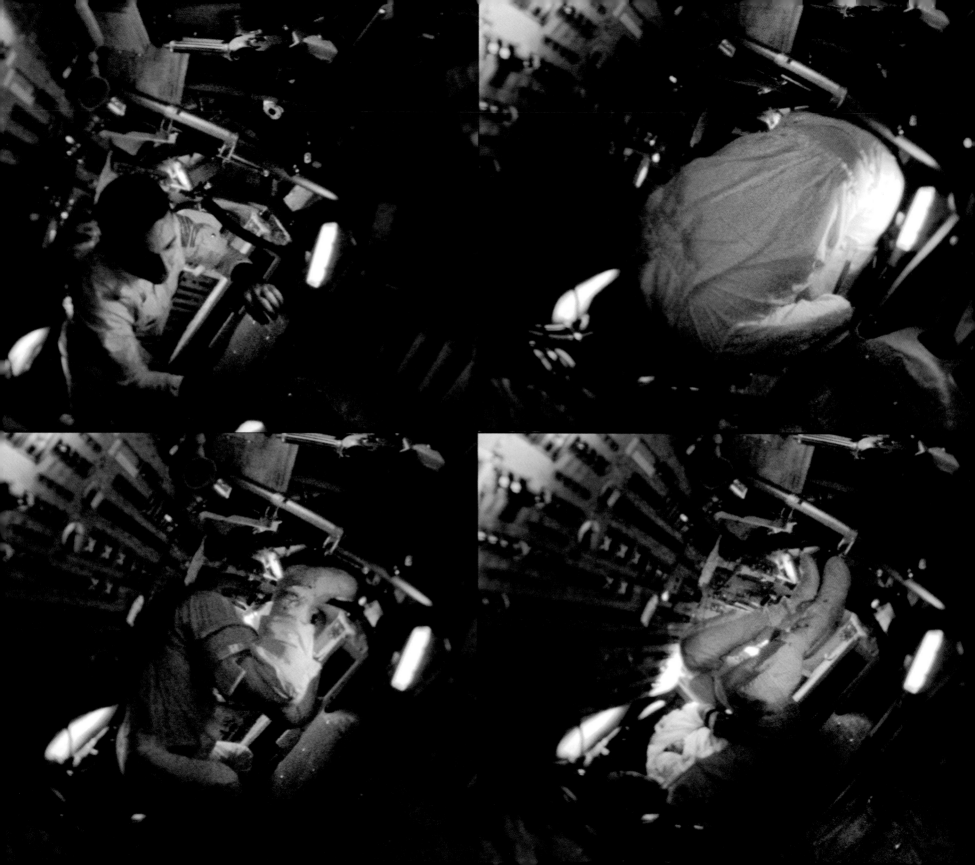

Inside Apollo 11 on the way to the moon, clockwise from above: Mike Collins in the command module's lower equipment bay; Neil Armstrong studies a photomap of the landing site; Aldrin inspects the lunar module.

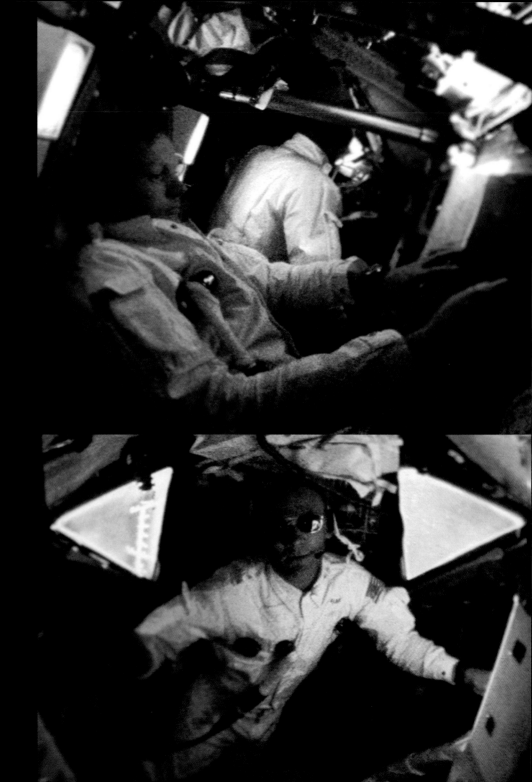

When we were about to leave on Apollo 11, [NASA administrator] Tom Paine flew down to the Cape, had dinner with us, and said, "Look, if you guys screw it up, don't worry about it, come on back, we'll give the three of you the next shot at another try at the landing." Which I thought was kind of unusual for him to say, but it was very nice, and it took a lot of the pressure off us, especially Neil.

—MIKE COLLINS

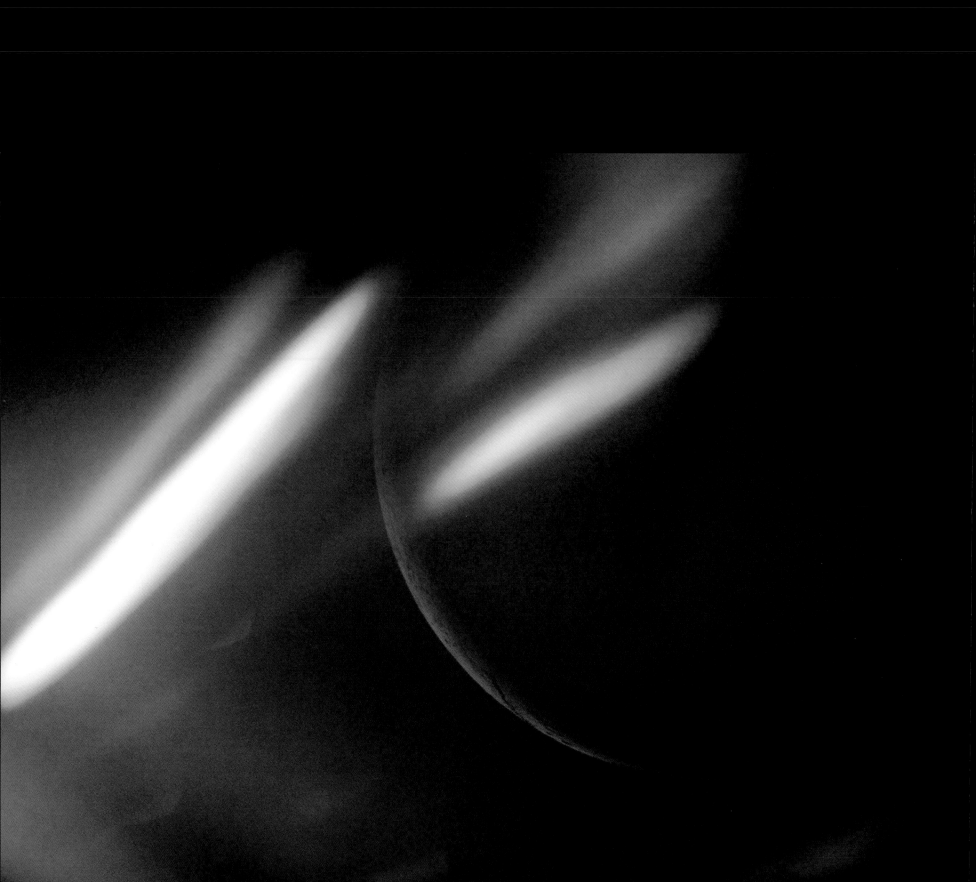

4

ANOTHER WORLD

It took a long time. And the fact that we couldn't look at the moon was kind of frustrating to me. I had envisioned—my vision of us would be peering out of the window, Jules Verne–style, watching the moon grow and grow and grow. But we never looked at it. And in fact, my first awareness of the *presence* of the moon was more of a hair-on-the-back-of-head type [of thing].

—BILL ANDERS

We went out to the moon, getting ready to go into lunar orbit. You know, whizzing in front of this big freight train, cruising through space. Turned around. . . . And I can remember—see, you couldn't see any stars. And then we went into the shadow of the moon. And *suddenly (snaps fingers)*—there were stars everywhere. God, there were stars—you couldn't hardly tell the *constellations*, because . . . even the dimmest stars suddenly popped out, and the constellations were somewhat confused. . . . Suddenly we get this *profusion* of stars. . . .

I can remember looking back, and suddenly becoming aware that here were all these stars, and yet there was this very sharp line. Absolutely no stars. *Total* blackness. And—*that* was the moon. . . . Anyway, I can remember literally feeling a little bit of a hair-on-the-back-of-the-head [sensation], thinking, "Are we falling into this black hole?" . . . And I remember feeling this kind of *eeeww* feeling, just like somebody would suddenly have if you go running through a tunnel blindfolded, you know, where you couldn't see the walls, even though you knew they were there. It was kind of an eerie feeling. And that was the first real awareness of the moon.

—BILL ANDERS

The ground told us that we'd lose radio communication at a certain time, down to the second, and by gosh, they were right. I was operating the computer and getting ready to light the engine to slow us down. . . . There's a button [on the computer] five seconds before you go, the PROCEED button. Essentially what it says is, Do you really want to make this maneuver, or don't you? It gives you a last chance to bug out of the deal. . . . Then, at the proper time, the engine lit off; it was being controlled by the computer. And after it lit off, it burned for a while, then it shut down on the display. It said, essentially, You're now in lunar orbit. Your orbit is so-and-so by so-and-so. I think it was 130 by 60 nautical miles, something like that. . . . Over the back side of the moon, of course, we saw the moon just 60 miles below us. . . . And . . . we were like three schoolkids. The most awe-inspiring sight. Looking at the back side of the moon . . . for the very first time.

—JIM LOVELL

I had some tapes with me, and I happened to be playing Sonny James's "How Great Thou Art" at this same time that we were drifting in behind the moon. And it was—I mean, the two together—here you have this unreal experience of watching the moon get bigger in magnitude, shrinking in phase, and you've got the tape recorder floating around with "How Great Thou Art." The first stanza talks about "When you view the universe and the wonders of God's handiwork." . . . It was just marvelous words from Earth as you're drifting in behind and watching the moon disappear, literally. Eventually you went into the darkness.

The experience of drifting in was so unreal. Then darkness. Then the burn. Then everything's okay. Everything checks, you're in a safe orbit. You go to a viewing attitude. And your first view of the moon almost knocks you out of the cockpit. So close. You could walk out and touch it. It's there. It's like the pictures, but you never really thought it would be so much like the pictures, so close, so real. I really wasn't prepared for that, even though I had studied those maps—I knew them exceptionally well. . . . I guess I never really had prepared myself for that initial view of the moon.

—STU ROOSA

On the way out, right at the very last, before we'd flown into the moon's shadow, John says, "Hey guys, look here." You could see a *little* [crescent], like one hundredth of a moon . . . just a little sliver. All the rest was dark. And pretty soon . . . we lost the sun, right within a second of when it was supposed to, it just got dark. . . . It was all the stars, and here was this big black area where there were no stars. . . . We kept going on, and finally . . . the Earth went down. We're just chugging along in the black, the three of us. . . . The clock's counting. . . . And even though we're getting close to the moon, we don't *see* the sonofabitch. . . . We say, "Jeez, I hope the ground's got this targeted right." *(laughs)* . . . No, we had faith. . . . And then, suddenly, you see it, right here, it just comes, *whooom!* Right out in front of us. Sliding right out below us. . . . And that's the first time we saw it.

—TOM STAFFORD

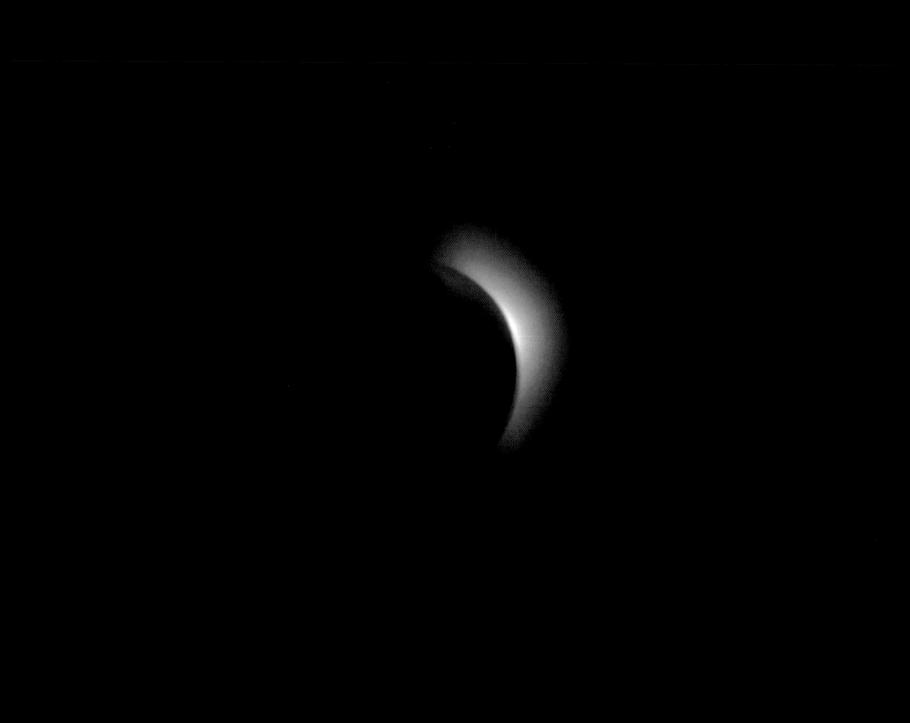

Bathed in the soft light of Earthshine, the moon is backlit by the glowing gases of the solar corona. Flying through the moon's shadow, the Apollo 11 astronauts had this view a few hours before entering lunar orbit.

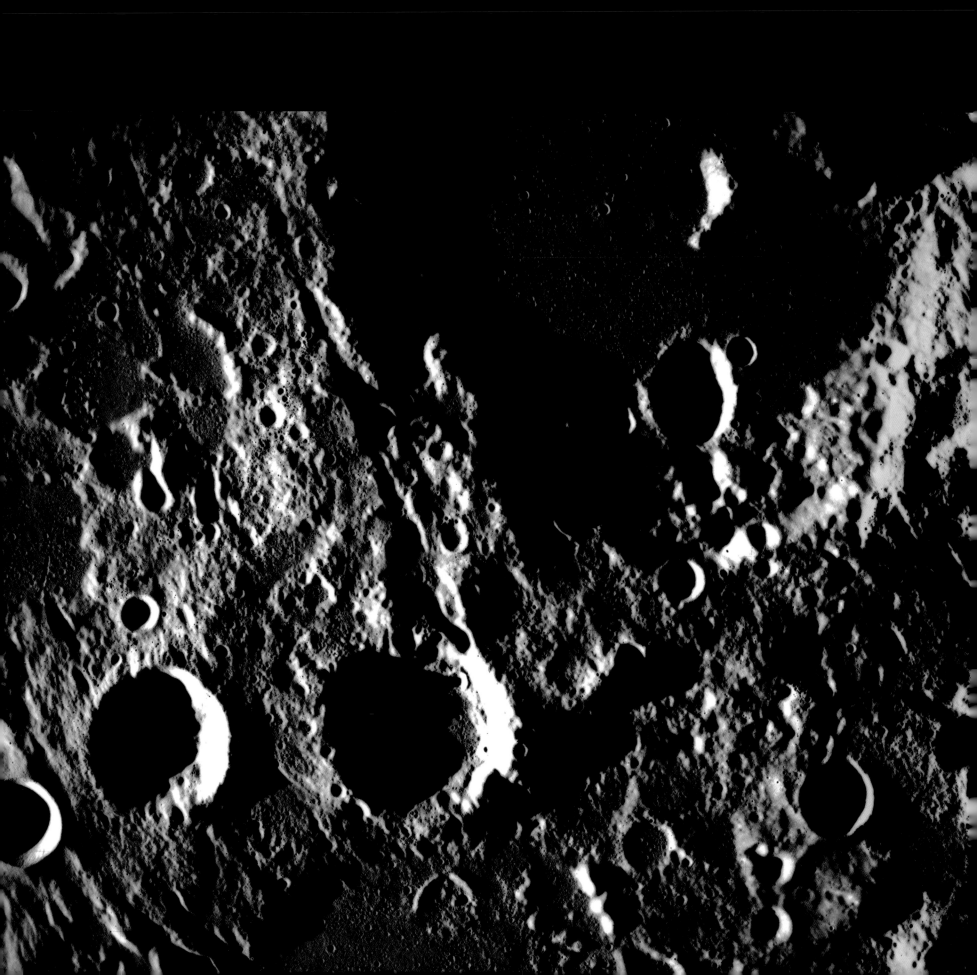

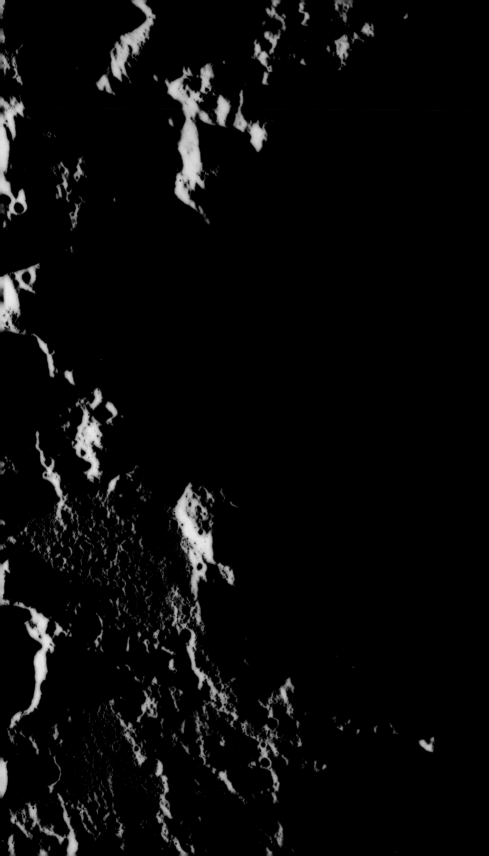

We made the [Lunar Orbit Insertion] maneuver and had just gotte[n] kind of organized. And by the way, we had all these lousy [haz[e] covered] windows. And I'm looking out . . . and I literally remembe[r] thinking, Oh, *shit.* Whatever that stuff is on the window, now it['s] really coming—it's *running down the window.* It looked like oil run[n]ning down the window . . . these rivulets of oil. And it was kind o[f] purply looking, because there was a purple haze on there. . . . S[o] thought, What in the hell is going on here? Geez, we'd had all th[e] trouble with these windows . . . and now I've got to look throug[h] this goddamn oil. And then . . . I refocused. And I thought I wa[s] looking at the window; I was actually looking at the moon. An[d] what it was was the long shadows [of] the very early lunar moun[n]tains. Now . . . that was a real thrill. I mean, to suddenly see thos[e] mountains . . . another world. . . .

—BILL ANDER[S]

If you look down at the moon, it doesn't look all that rugged unt[il] you get over near where the terminator [the boundary between da[y] and night] is. And then it looks like, my God, these sharp ridges—[] I mean, you think it's the worst looking part. And I remember th[e] first day we got to the moon looking out there, and I thought, m[y] God, look at that part of the moon—there's the *bad* part of th[e] moon. And then the next day I said, Gee, this is *still* the bad part o[f] the moon. It was only when we got back up to lunar orbit [after th[e] moonwalks] that I realized that all of the moon's that way; it's onl[y] when it gets in these long shadows, and the shadows start makin[g] these jagged lines, that you can see how rugged the moon is. Whe[n] you're like, at noon, hell, everything just looks like flat craters.

—ALAN BEAN[]

No matter how closely you looked, it was crater upon crater. You['d] get the monocular out, and you'd see just even more craters. So you just imagined they were right down to microscopic little [craters]. What you had down there was a big heap of crater upon crater.

—BILL ANDERS[]

A rugged moonscape seen from Apollo 16.

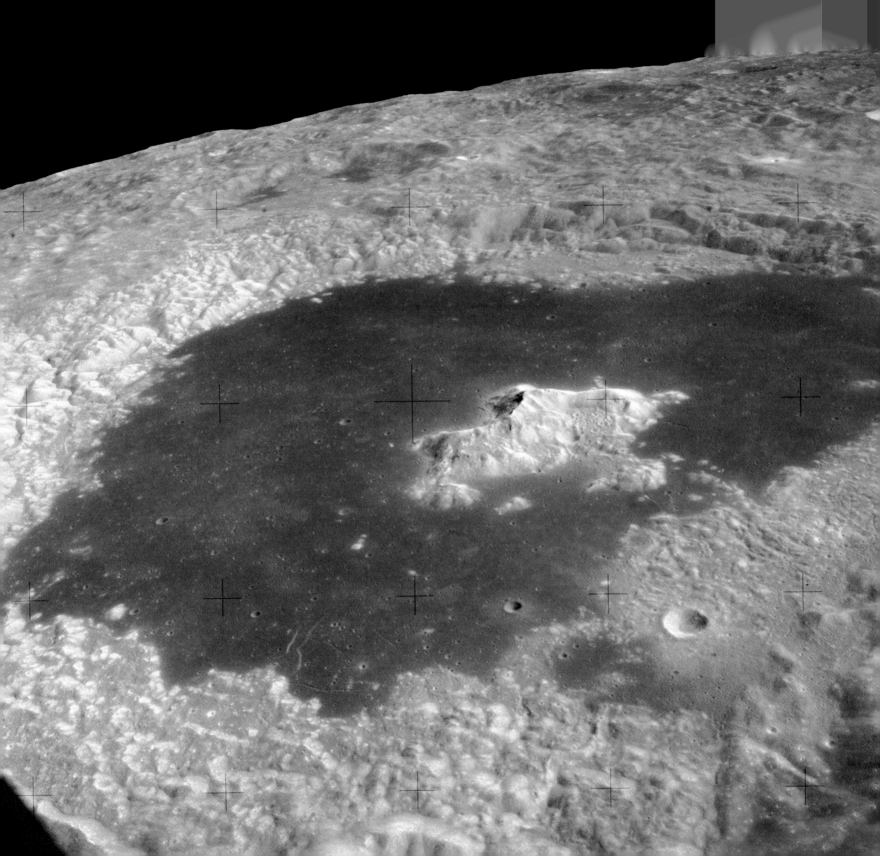

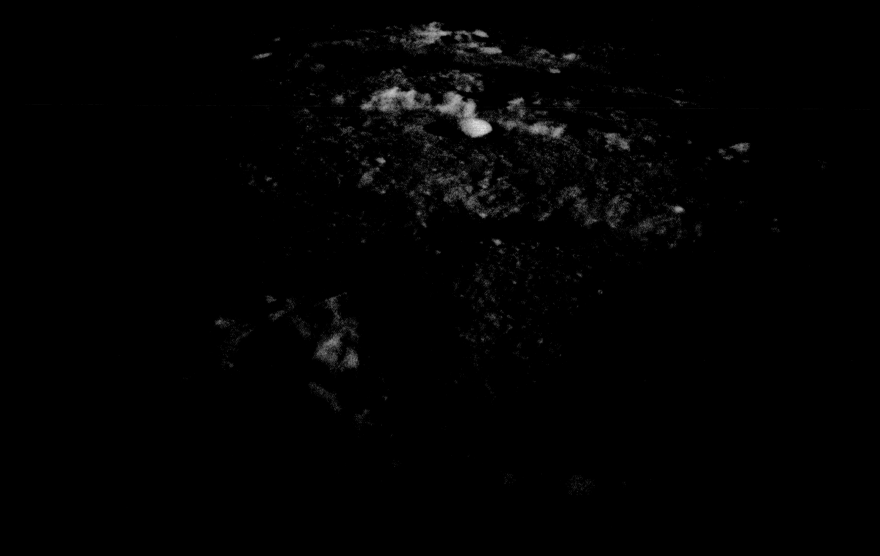

The back side is more mountainous than the front side . . . someone said it was like papier-mâché. Well, right, it's all shades of gray. There is no color.

—JIM LOVELL

The biggest problem, going around the moon . . . is knowing, instantaneously, when you look out the window, Where am I looking? You can't keep track of where you are. You're always interrupted to do something else or whatever. You've got to tell yourself ahead of time to figure out where you are, so that you know where to start looking for the next landmark, the next thing that's coming up.

—RON EVANS

Seeing the moon in Earthshine is like flying over snow-covered terrain . . . with a bright moon and totally clear skies. . . . You get this magic terrain. And you can see relief, and you can see things. But it has this sameness, this uniformity in color that comes with snow scenes. And that's the way the moon is . . . except you can see more detail, because Earthshine is so much brighter than moonshine. . . . It's just an absolutely extraordinary sight.

—KEN MATTINGLY

Above: *The moon in Earthshine, photographed from the Apollo 17 command module.* Opposite: *The giant, dark-floored farside crater Tsiolkovsky, seen from the Apollo 15 lunar module.*

There's no doubt in your mind that it was smaller [than the Earth], once you got there. You're only sixty miles above it, and it's all curved away from you.

—PETE CONRAD

[There are] changes in the color of the moon as you traverse from sunrise to sunset. [At] sunset . . . you think the moon is brownish. . . . Then the brownish gets lighter brown, and pretty soon you get over to high noon. . . . And you look out there and it's bright, *bright*, really, really *bright*. . . . You look around—Where in the world are my sunglasses? And the damn things have floated off somewhere *(laughs)*.

—RON EVANS

The thing that impressed me about the moon is how different regions are so different. Now, they all look the same to the geophysicist. But they sure look different to the people that [are orbiting] close in. Every little crater's got a different characteristic. . . . And it's sure hard to pick out till you get down there and really look at 'em. And I expect that when we get back to the moon, we'll learn a whole lot more things, that we'll say, Well, that sure was obvious when you thought about it. . . .

—JOHN YOUNG

We were *there*. I mean, if things didn't work going [to the moon], we were going to get a free ride home, on a free-return trajectory. Maybe the reentry wouldn't be perfect, but at least we'd have a shot at it. Once that rocket worked [and] got us into lunar orbit, then it *had* to work again, or we were stuck.

—BILL ANDERS

If this engine didn't light [for Transearth Injection], you were in trouble. But we never gave it a second thought. . . . Well, you live with that stuff for so long, you've got to have a certain faith in the equipment. . . . If you always have second thoughts, you would *never* get in that program. You must *assume* that everything is going to work right. . . . If you worried about it ahead of time, you'd be a basket case if something really went wrong.

—JIM LOVELL

An Apollo 15 view of the Sea of Ingenuity.

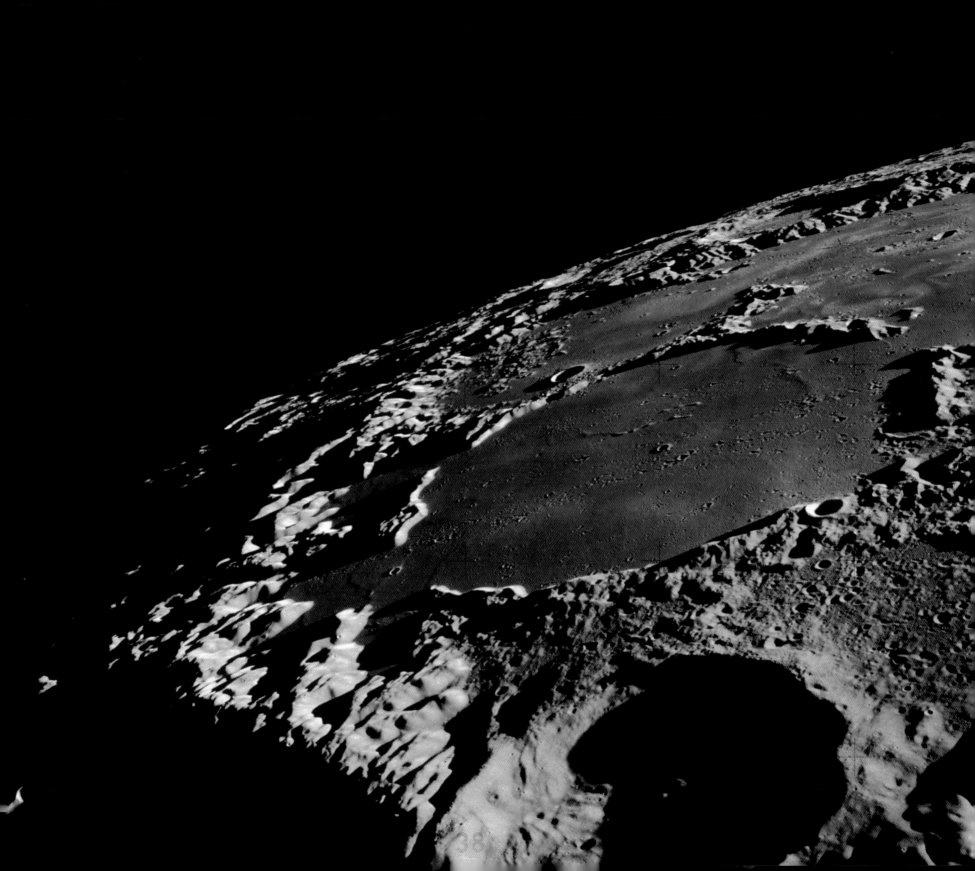

On our flight, most of the back side was illuminated. Unlike Arthur Clarke's *2001*, this back side was very beat up. It was rough. . . . It came across very rapidly that this was *not* a very interesting place. . . . Just beat-up, dirty beach sand. *Pulverized.* It looked like a war zone. . . .

Suddenly, *wow*—in contrast to the moon being *less* than we expected, suddenly this Earth jumps up. Which is, one, a lot prettier than we expected. Two, it's in contrast to the moon, which amplifies its beauty. You've got to remember, we were 240,000 miles away from home. And so, *that's home!*. . . It's very delicate. It reminded me of a Christmas tree ornament. Very fragile, delicate. And you could imagine that we only live in that tiny little skin around the outside. . . . The only color you could see in the whole universe. Everything else was black or white or gray. But here was the only color: blue. God, that blue looked pretty. . . .

That was the most beautiful thing I'd ever seen. . . . The Earthrise. Totally unanticipated. Because we were being trained to go *to the moon*. . . . We were trained to *get there*. So, *getting there* was the big event. . . . It wasn't "going to the moon and looking back at the Earth." I never even thought about that! . . . In lunar orbit, it occurred to me that, here we are, all the way up there at the moon, and we're studying this thing, and it's really the Earth as seen from the moon that's the most interesting aspect of this flight.

—BILL ANDERS

The fact that you can put your thumb up to the window of the spacecraft and completely put the Earth behind your thumb is a concept that gives you the insignificance of your own existence with respect to the universe. . . . You have to think about hiding the entire Earth. Everything that you've ever done. All the people you knew. Every place you've been. Continents. All the major confrontations that you had. You know, the various wars that were going on at the time. The problems at home, the dissension. . . . Nineteen sixty-eight was a very bad year. The Vietnam War was going on. The Democratic convention here in Chicago was a disaster. It was not a good year. . . . And it turns out that it really made you feel humble, because everything shrunk in size.

—JIM LOVELL

The first Earthrise, photographed in black and white, then in color, by Bill Anders from Apollo 8.

You know, I would like to say that I was in favor of that [TV show from lunar orbit]. . . . I didn't want to take television because I thought it would interfere with our mission, and the weight and the complexity of it. We took it, and it was one of the smartest things we did, because it brought the public into the mission the way it should have been brought. But . . . *(laughs)* I didn't want the damn thing around!

Once we got there and saw the moon, I recognized this was extremely the right thing to do. Because the moon was so desolate. . . . I think the moon resembled what the Earth must've looked like before there was life. Or what it could look like after an all-out nuclear war. So that was sobering.

—FRANK BORMAN

To me [the Genesis reading] was that we're trying to say something sort of fundamental. Something that will stop and grab people's guts and say, hey this isn't just a little whistle-around-the-Earth space shot; this is man's . . . first step away from his home planet. I mean, we're talking about a second Genesis, if you will.

—BILL ANDERS

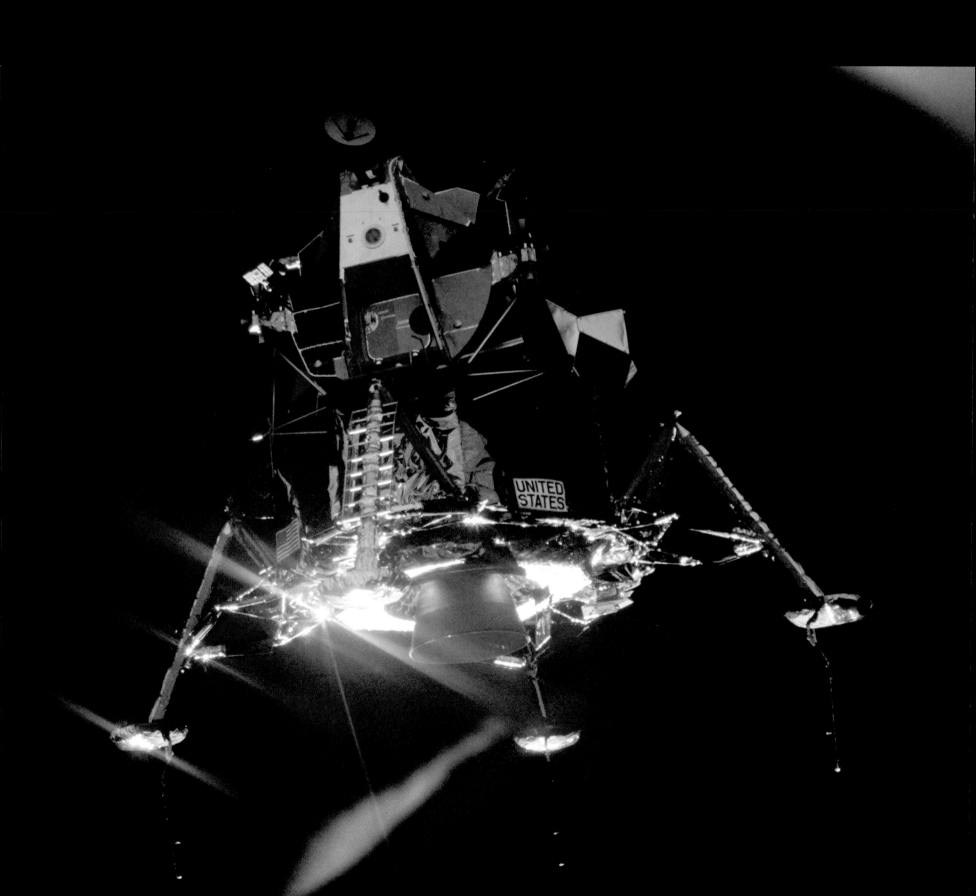

5

LANDING

I've often said that my instinct—not a carefully reasoned statistical study, but my instincts—told me that we had a 90 percent chance of a safe return and a 50 percent chance of a safe landing.

—NEIL ARMSTRONG

landing the lunar module was . . . the kind of thing where you know that you only have one chance—no two chances, one chance. Everything has to go right. So that puts you right up on the edge of performance. . . . There's an old saying in the program, I don't know whether any of the other guys have mentioned it: "Get ahead and stay ahead." . . . Always stay ahead. . . . So in a lunar landing, it's really thinking ahead. It's planning ahead. . . . Because if you get surprised, it's going to take away from your time and your mental process. And if you're ahead, you can absorb that. . . .

So you're thinking all the things that you should do, and all the responses to emergencies—you don't get into specific emergencies, but you're just running your [mental] computer as fast as you can run it. . . . And if you get into an emergency situation, things happen so fast, you have such a short period of time, there's no margin. You know—thirty seconds, or whatever it is. . . . You cannot afford to make a mistake. . . . Not so much consciously, but subconsciously, you have all your memory banks running. You're focused entirely on the job, but you're also paying attention.

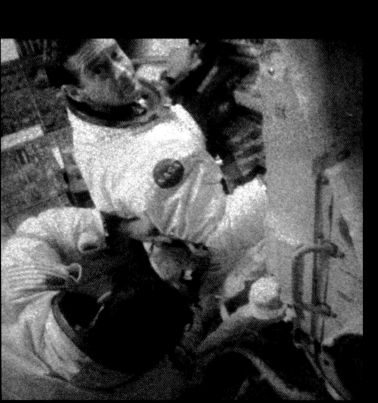

As an example. I'm holding onto two [control] handles, and there are a number of buttons in front of me. Now there's a blue button to turn the engine off, and there's a red button to abort. I don't want to push the red button. . . . So when you go into the landing, part of your computer in your mind is concentrating on those three buttons so you don't screw up. In addition to that, you're concentrating on the flying, and you're concentrating on listening to Jim and looking out the window, and the [trajectory]. . . . What if all of a sudden we lose [communications]? What if I can't hear Jim? Then I gotta know what he's doing. So I have to make sure that without him, I can still do the job.

So you play all that [in your mind]. 'Cause in the simulations, they've done that to you. They've done all these things to you. . . . That's the beauty of the simulations. . . . In the simulator, you can say, I'm not going to listen to Jim this time. I'm just going to go do it. But in the real world, you have to put everything in the [mental] computer and run it at the same time. So the mental challenge is enormous! I mean, you don't have to focus on all that, but you better. Because if you don't, number one, you could screw up anyway, and number two, if you have a problem, it diverts your attention, and if you're not in parallel processing—and that's what it is, parallel processing all that stuff—if you're not processing everything in parallel at full speed, you're liable to miss something. And if you miss it, either you're dead real quick or you blew the mission. There's no recovery. And you know that, going in. There's one chance, and you've got two or three minutes. One chance and that's all. So, boy, you really tune up for that.

For that reason, flying the lunar module is a very demanding task. It's the toughest flying job—and I've flown a lot of stuff—the toughest flying job I've ever had.

—DAVE SCOTT

Left: *Inside the Apollo 16 lunar module,* Orion, *John Young helps Charlie Duke suit up before undocking.* Opposite: Orion *flies free, seen from the command module,* Casper.

It was a beautiful airplane. I think in some of the failure modes . . . that it was very difficult to fly. If you had total three-axis failure of the autopilot, the only way that we could ever get the simulator down was for Ed to do the yaw and the roll, and I would handle the pitch and the throttle. And, you know, we made some hairy-looking approaches [in the simulator]. . . . Because, you see *(gesturing with his hands),* here the thing's sitting like that [on its engine exhaust]; it's literally balanced. You just move it one way and it's just going to keep going on over, unless you stop it. And so you've done that maneuver, and it starts picking up speed, and then you've got to do the exact opposite. So it was a real spastic thing. And we hoped, obviously, that we'd never have a failure in three axes at once, because one man couldn't do it without losing it.

—ALAN SHEPARD

You had to get it down or you didn't get another chance. . . . I mean, at the [aircraft] carrier, you could always go around.

—PETE CONRAD

We all adapt our relationship to the passing of time, depending on the circumstances. A batter looking at a ball, you know, reads the spin, tries to figure out where the ball's going to go, and takes a swing, all in a matter of less than a second. On the other hand, a lot of things, you take very long deliberations, take a hundred or a thousand times as long to make a decision of similar import. And I think it's just human nature that you adapt your appreciation of the timescale to meet the circumstances. In the lunar landing, when you have not too much fuel and not too much time, you adapt your sense of time to that which is available. And if you're properly trained, and you have enough practice, you can get a confidence that you can do what's required in the time available, and you really don't worry about the pressure of time.

—NEIL ARMSTRONG

Pete and I get into the lunar module, getting ready to go down to the moon. And Dick's job was to put in the probe and drogue and stuff. It took quite a while for him to do that, and we were getting stuff ready, and trying to close out the lunar module . . . and all of a sudden Dick said, "Well, I'm getting ready to close the hatch now." So we looked up at him, and—I can still see him up there getting ready to close the hatch—and we didn't really say anything; I can't remember if we said, "See you in a couple of days," or didn't say anything, you know. I personally was wondering, "Will I ever see this guy again? Wonder what's going to happen to us? I hope I see him in a few days." But we never said anything, we kind of looked at each other, you know, and had, really, I felt, really loving feelings between us, yet I don't remember saying anything, except something light, like, "Well, see you in a few days." . . . It seemed that we had a lot of unspoken thoughts there, at least I did. You know, when you don't speak things, then you don't know what the other people are thinking.

—ALAN BEAN

I tell you what, I envied them. I wish to hell I could have gone with them, but there was no way for that to happen. I think Pete and Al kind of felt the same way.

—DICK GORDON

Inside the Apollo 12 command module, Yankee Clipper, *Dick Gordon photographs the departing lunar module,* Intrepid, *carrying Pete Conrad and Alan Bean.*

I thought in a gross sense that the lunar module was a much better flying machine than I expected, and it was really easier to fly than any of our simulations. . . . Talking about a fairly few seconds of control here, and it's difficult to draw fine judgments on that limited experience. But I felt good about the flying qualities of the machine. . . .

There was a substantial distraction [from the computer alarms]—because one wonders, Is this something serious that I have to worry about? . . . It surely diverted my attention. And normally in that particular phase, I would have been selecting landmarks and trying to identify precisely where I was going to go and where that related to our intended landing spot. And that's the part that I missed, that I didn't get to do. . . . All our landmarks that we used for identifying our target were upstream. They were east of the landing area. We'd passed those. . . .

That [football-field sized, boulder-strewn crater] was clearly a very desirable location [for the geologists], because there was a lot of action there, there were a lot of things that people would be interested in, and that looked like really an ideal place *(laughs)* to go if you could do it safely. But I didn't have that much courage *(laughs)*. . . .

Could be a lot of things [that would make it risky to abort the landing]. Could be that you don't get an engine fire with your ascent engine. You don't get clean separation. After all, that's a whole 'nother test program, doing an abort sequence at low altitude for the first time, so in that sense, with a landing, at least, you're not tearing the vehicle apart, not changing engines in midstream. The landing gear's already down and locked. You have a lot of things going for you, and I would agree with [flight director] Gene Kranz that, other things being equal, you might tend to push in the land direction rather than the abort direction as a relative risk judgment.

It [blowing dust] wasn't a surprise. We expected that there would be some kind of effect from the rocket exhaust on the surface. But not knowing precisely what the surface was, and not knowing what the deflected exhaust characteristics [would be], we didn't know what to simulate and how to simulate it. . . .

We were pretty close, pretty low, and I was close enough that, I thought if the engine quit [because of running out of fuel], we're alright—you just fall into the moon. So, when you get down to a certain level, you don't care if the engine quits. Just like an airplane *(laughs)*. I don't think I was thinking about that, except subconsciously. But there's a point at which you are worried, and that's when it's very difficult,' or maybe impossible, to abort. And once you get below that altitude, and with the velocity low enough that you aren't going to tip over when you come down, then you're the same as free—home free.

—NEIL ARMSTRONG

There was that moment, right after we touched down, when . . . we just kind of looked at each other and—I'm not sure how it happened, a slap on the back, or whatever—but there was that, just, little moment of, Hey—we made it.

—BUZZ ALDRIN

The view from the descending Apollo 11 lunar module, Eagle, at an altitude of (top left) 7,000 feet, (top right) 200 feet, (bottom left) 75 feet, and (bottom right) 7 feet. These frames were recorded by the onboard 16mm movie camera.

To a certain extent, you console yourself—it's just another simulation, and you've been through all sorts of these simulated problems. . . . Well, you know better, but it's a good trick to calm yourself down, to say, It's just like a simulation.

—ED MITCHELL

I think I mentioned . . . the surprise we had when we looked up at six thousand feet above the surface to find this mountain on our left went another seven thousand feet above us. We'd never had that view out of the simulator window. . . . Fortunately, Hadley Rille was very obvious out in front of us, and that reassured us that we'd probably come to the right place.

—JIM IRWIN

I found it very convenient not to look out too much. Because it was very distracting to see all the craters, and see the moon, and see you were coming down. You know, you're really doing this. If I'd look inside, it seemed a lot like the simulator. So, what I would do was look out, and then when I would get excited, or (laughs) full up with information, you know, I'd say, Quit doing that and concentrate on what you're doing. Then I do it, but then I'd say, Well, I don't want to just miss the whole thing, either. I want to do the job, but I want to not miss the show. So it was a constant look in and do the job, and then look out and try not to look out too long.

—ALAN BEAN

[The dust] was very confusing. . . . It's very confusing, looking at that stuff going out laterally, and you're not really sure what you're doing. So you had to check the gauges, or look for rocks sticking up through it, or whatever. And the window's not very big. And the closer we got, the worse it got, obviously. . . . And it got heavier and heavier to the point where, if you look at our movies, just before touchdown there, no way you can see through it to the ground.

—PETE CONRAD

Damn right I was watching where we were going. If anything went wrong, I wanted to be oriented. Fast scan pattern—I was watching inside, but I was keeping things posted as to where we were. I wanted to be in a position so that if anything happened—Al's controls went out, or anything—I knew how to handle it from that point. I was ready to go ahead and land if necessary, or whatever. . . .

The both of us knew we were going to land. Even when the landing radar didn't come in, it was pretty certain to pitch over, and if we were in any sort of position, we'd've landed anyhow. We're not going to get down to eighteen thousand feet and not take a look. Regardless of what Houston says.

—ED MITCHELL

Ed said [later], "What were you going to do?" I said, "You'll never know" (laughs). . . . We had an altitude limit, which we couldn't go below unless we had landing radar locked in and feeding into the computer. And we were getting awfully close to that. . . . Oh, I'm sure if we had not had it in, and I'd pitched over and gone down and landed, you know, he wouldn't have said, "Don't, don't!" (laughs)

—ALAN SHEPARD

To shut down and drop to the surface was a real relief.

—ED MITCHELL

The view from the descending Apollo 12 lunar module, Intrepid, 650 feet above the Ocean of Storms.

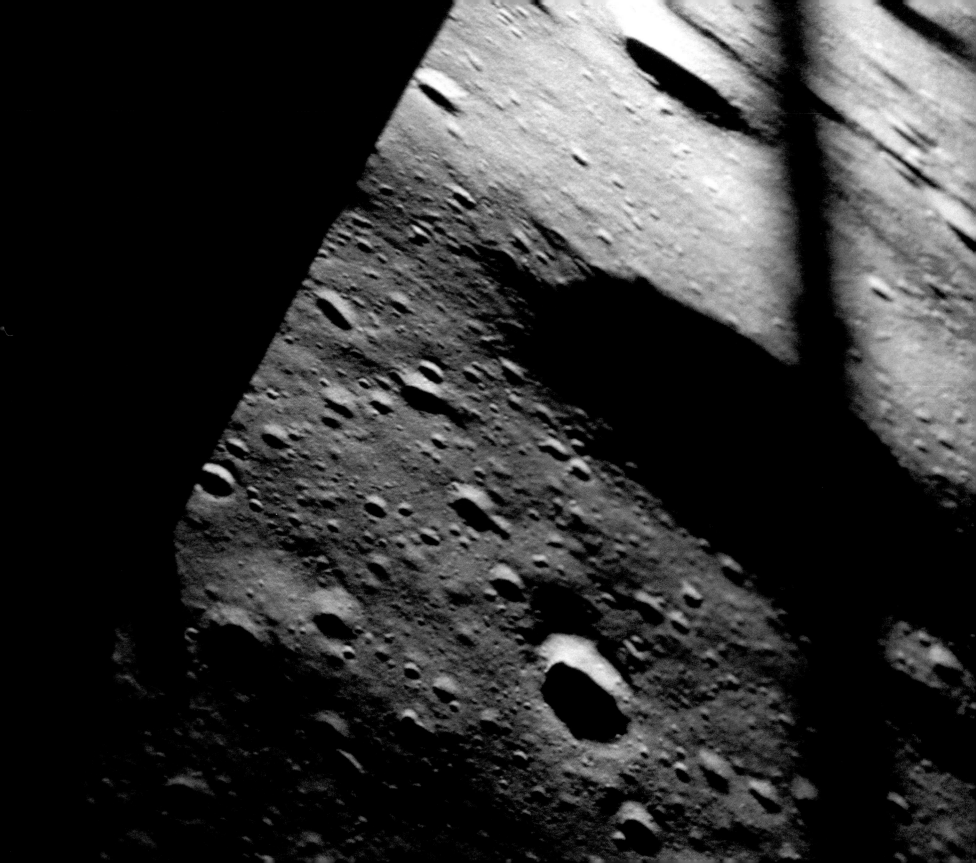

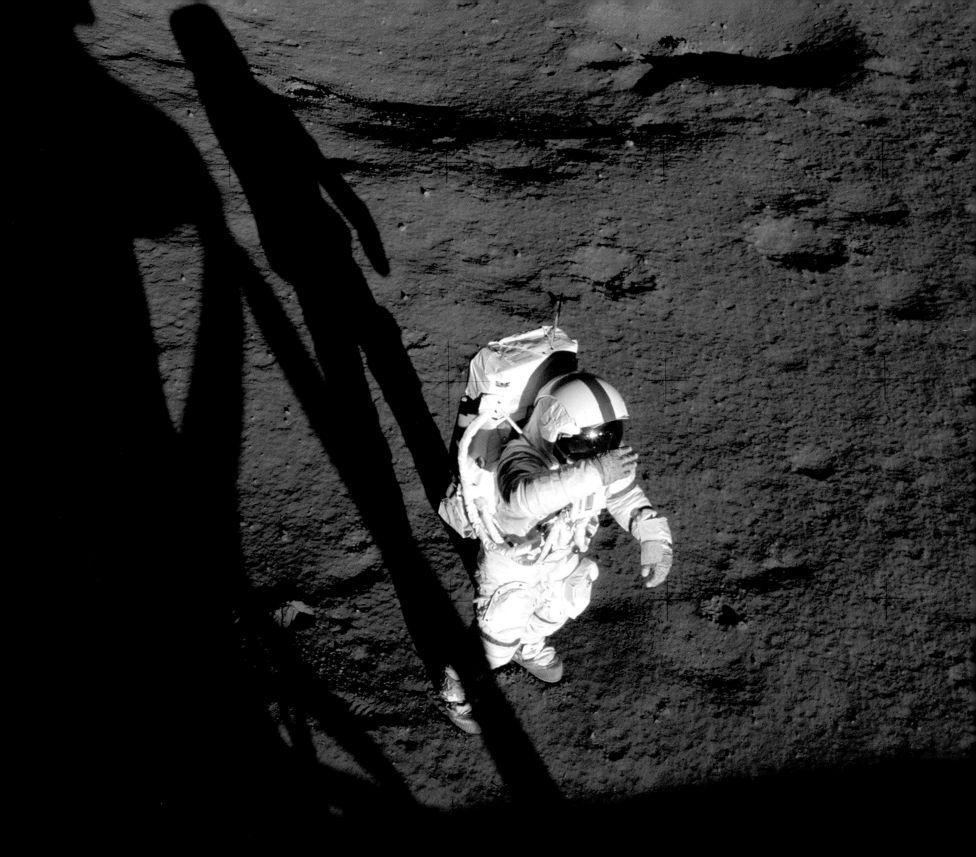

It was the irony. It was the same irony that caused me to think, pause, and just inwardly chuckle, just momentarily, that, God, here are two guys further away from home . . . than two guys had ever been, but there are more people watching us than anybody else has ever watched two people before in history.

I found it to be a very comfortable place to be, for the plain and simple reason that we weren't in flight. And I had not given that consideration. I had always thought, you know, you're at the moon, and your 250,000 miles from home, you're going to be lonely, and you've really got your neck out. But what I did not put in the equation was that you were going to have landed. And the fact that you had landed on the *moon* didn't have anything to do with it. The fact that you were on solid ground and you were not flying was something I hadn't taken into the equation, [and] that was a very comfortable and safe feeling.

—PETE CONRAD

I saw just the opposite [of Conrad]. I felt that from my history with cars and airplanes and things, that once you're on the way in your car you're probably going to get there, maybe. The time you really don't get there is when you go out to start it up in the morning and your battery's dead. I've had less problems with cars driving along—can't even think of one, or—maybe one or two—but I've had a lot over the years [where] you get in and it didn't work. Or an airplane wouldn't start. A lot of those with airplanes that don't start—none ever in flight.

So when we stopped on the moon, I was glad to be there, but at the same time, I felt a lot more vulnerable down there. I never felt as safe on the moon as I did the minute we got going to orbital speed. Somehow, when we got orbital speed around the moon, I said, We're going fast again. If we can just get going a little bit faster, we'll be back home.

—ALAN BEAN

It was a natural thing to me to be there. . . . I didn't feel out of place. It was acceptable. . . . It was spectacular, it was different, but it was accepted. In my mind, I accepted that. Okay, we're on the moon. That's terrific. Great. Now, let's go do some geology.

—DAVE SCOTT

It was probably the most hostile environment I'd ever been in in a flying situation. And yet I felt more at peace and more serene. . . . You didn't feel *(pauses)* fearful of the hostile environment. . . . There was a serenity and a peacefulness about the moon. There was a great peace inside. There was a sense of security in the fact that we were here and the mission was going to get done. A lot of confidence in our ability. We had trained for two and a half years—*we knew what to do.* . . . There was a feeling of belonging. I felt familiar. . . . There was a sense of familiarity—there's Stone; there's North Ray; there's Smokey; there's Dot crater. I'd looked at those photographs for a thousand times! And here we were! We're right here on this photograph. And so I felt right at home . . . I believe because the training was so thorough and our preflight planning and study of the landing site made it *(pauses)* preconditioned. Now, if the lunar module went arcing off somewhere and landed forty miles north or forty miles west, I think it would've been a lot eerier.

—CHARLIE DUKE

I'd been eating, sleeping, and breathing it for seven years. I was supposed to be there then, and I was there, and it was the right place to be, and I didn't think it was strange, I didn't think it was awe-inspiring or anything else. It was where I was supposed to be.

—PETE CONRAD

The Apollo 14 lunar module, Antares, *rests on the moon's Fra Mauro highlands during the mission's first moonwalk.*

I felt all along that I was one of the more fearful astronauts, one of the more cautious ones. I would be more nervous about things than someone else, about the same situation. It took more personal effort to overcome my natural fear. . . .

I think you do it all along in the training. You think about these things every day when you're at meetings, and they're talking about what happens if one engine fails or . . . what you do in an abort mode. . . . And so you're constantly thinking about it. You continue to think about it all the time, even after you get back. But you're constantly weighing the risk versus the gain. . . . [You] try not to think about it. . . . If I found myself thinking about it—try to make all the decisions about it beforehand, before you go. What you're going to do, whether the risk is worth the gain, all that stuff, way beforehand. And then when you're there, just try to carry out the routine of it. Try not to get off the plan. . . . If you notice yourself thinking about the fact that it's dangerous, say, I need to quit thinking about that. And now I'm supposed to be digging a rock up or something. . . .

It's just outside of consciousness. For example, when we were sleeping, there was a change in pitch in our lunar module. It was the fans for the environmental control system. . . . The minute that changed tune, both Pete and I were up and looking around, just like (snaps finger) that. . . .

You know that if the suit breaks, you're going to die, or that your backpack could quit working. You know, you're outside in the vacuum in this little suit, and although the suit's worked lots of times, you know it could quit working. . . . And you're surrounded by death, you know, you're surrounded by this vacuum when you're in your suit. When you're in the spacecraft, the vacuum and death is out a window. You know, it's over there. But when you're in your suit, then you're in it. It's a whole different thing. I guess it's like . . . in your house, looking out a window at an unruly mob or something. That's one thing. The other one is to be standing in the middle of it . . . you know, where they're looking for you. If you're in the house and they're looking for you, you've got some protection. If you're outside among them and they're looking for you, well then, it's a little more dangerous. . . . That's why I think Pete was singing when he got outside. Just happy to be there. Also, I think singing takes away some of the tension and fear that's present. . . . I think there's a lot of tension up there—more than anybody talks about, because of the Right Stuff syndrome, you know.

—ALAN BEAN

I just didn't think about it. "Man, this is neat," is really what was going through my mind. "Man everything's working great." We had a few problems, but we were expecting a few. So, you know, I didn't sit back and say, Man, I'm on the moon, 250,000 miles from Earth, and if this zipper breaks or if this window pops out, we've had it. . . . You just have so much confidence in the equipment and in your ability and in the familiarity that the training had bred into you. . . . In a sense, it's faith in yourself.

—CHARLIE DUKE

You've got to get rid of that before you go. If you're still afraid to be up there, then you shouldn't be up there.

—ED MITCHELL

Alan Bean takes his first step onto the moon's Ocean of Storms.

You can see pictures, and you can appreciate some from pictures. But pictures is *different* from being there.

—NEIL ARMSTRONG

There's a *starkness*. There's a precision, and yet there's a disorder. . . . It's the precision of dust and rocks, and definitive horizon.

—BUZZ ALDRIN

It's just craters and craters and gray dust. And yet, you know, it's like the Grand Canyon. The Grand Canyon is just rock and a big gash in the ground. But it's still beautiful. It's the same with the lunar surface. . . . The lighting's incredible, the contrasts. It's something you've never seen before. In that stark sunlight, it's either black or it's gray or it's dazzling sunlight. Nothing muted about it. . . .

There was nothing different than what we had trained to believe, just more of it. I mean, it was exaggerated. The simulation—we came to expect it to look like the simulation. Well, it did—but it was real. So there wasn't anything you should've been terribly amazed at, except you were amazed at it. Just because of the starkness of it, the reality of it, the beauty of it, the fact that here you were after all this time.

—ED MITCHELL

It had a majestic feeling about it. And one says this after talking about how it's dusty, it's gray, nothing's growing, nothing of any real beauty. But yet, take it all together with the vastness of it, the sense of history, the boulders, and the elevations we had on our flight and certainly some of the other flights—Hadley, for example— it really is majestic, in the sense of a desolate mountain desert type of a setting.

—ALAN SHEPARD

it was the ultimate desert.

—JIM IRWIN

Oh, the beauty! The spectacular beauty. . . . Oh, yeah, that's, that is, to coin a phrase, mind-boggling. . . . It's absolutely mind-boggling, because you cannot believe that it is really that spectacular. . . . I didn't expect the beauty of it. That's the one thing everybody talks about, and you can't appreciate it until you get there and see it.

—DAVE SCOTT

The disappearing of the horizon—it is distinct. There's no haze, no nothing obscuring it, and that makes distant objects appear clear, very clear. And then there's the rock, and then there's nothing. You look out and you see that. Whether you realize it or not, you're looking at the edge of a ball, and you're on it. And that sensation was clear as *distinctly* different than on the Earth. It *is* curving away. Not that it just grabs you immediately that you're on the knoll of a hill—it's not that; it's more than that, but a little bit of intellect and a little bit of everything else says, Gee, this is really obvious that this is a sphere that we're on and walking on.

—BUZZ ALDRIN

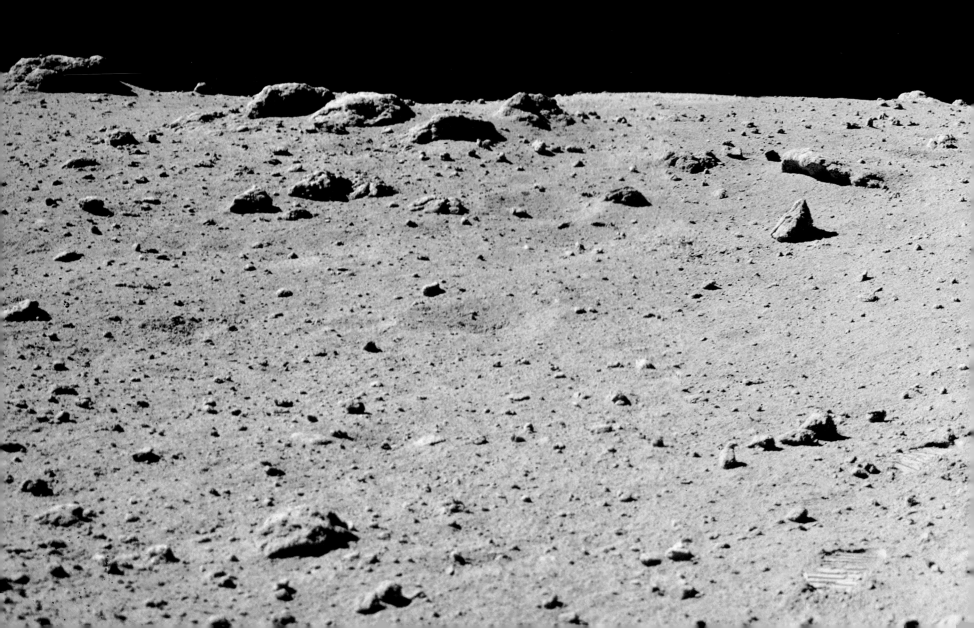

That blackness you see in those photographs is just black black. . . . You feel like you can go over there and it's a black velvet screen—it's just a backdrop for this stage deal we did. So that's the feeling of space, that you can just reach out and touch it. And yet there's nothing there.

—CHARLIE DUKE

Most people can't comprehend a black sky except at night. But we can comprehend a black sky in the daytime. Because on the moon you have a black sky. And it's very different from a blue sky. I mean, when the surface of the moon is illuminated, and it's bright, and there are shadows and contrasts, et cetera, and then you see the ridgelines, and above that is a black sky—that is a whole new thing for . . . the mind to handle. . . . And that's why we say, Ooh, this is spectacular.

—DAVE SCOTT

Near the rim of North Ray crater, Charlie Duke works at the rover during Apollo 16's third moonwalk.

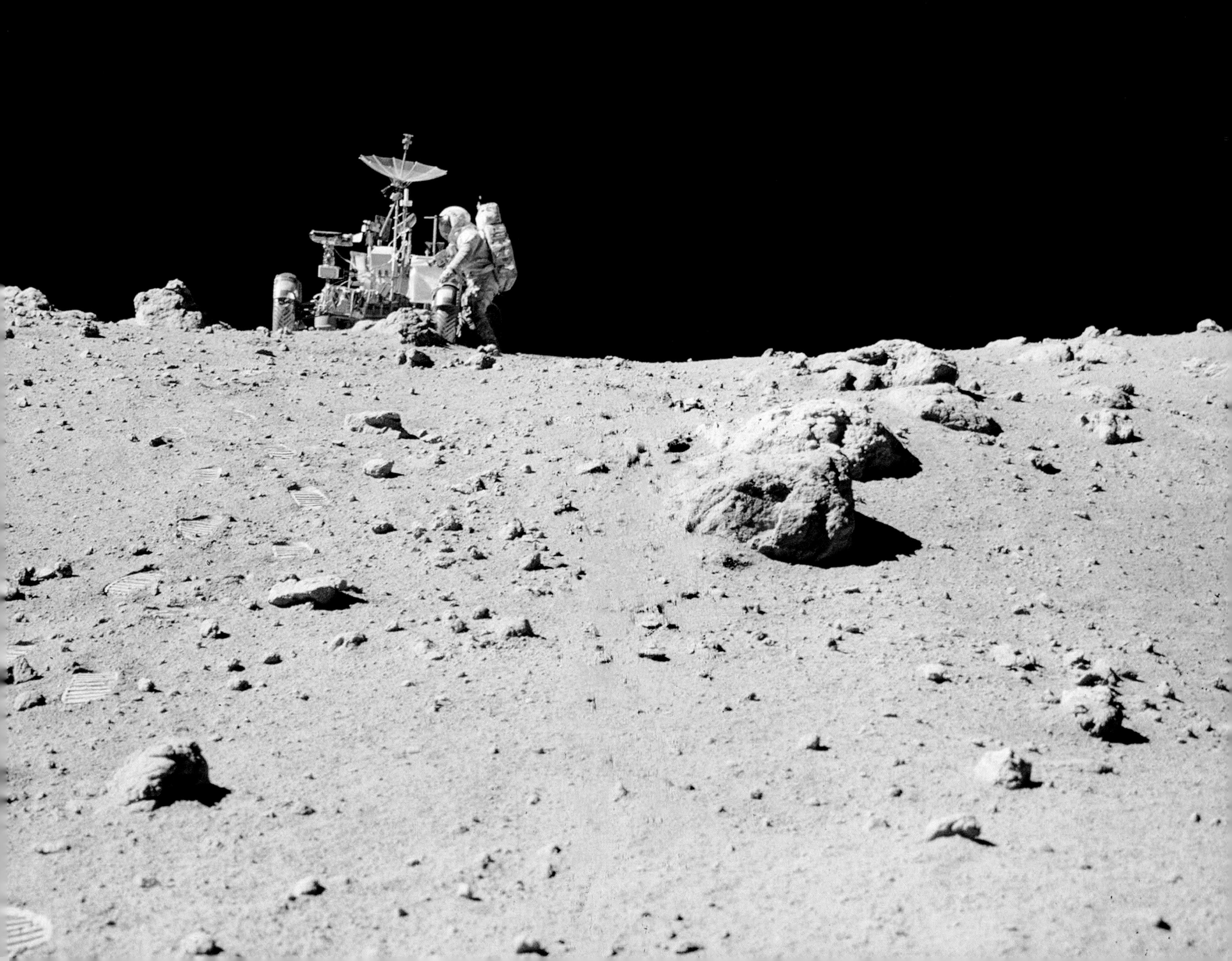

They [the mountains] looked big, but not as big as they were. Interestingly enough. And—they were enormous. They were huge. But I expected them, frankly, to look bigger in the scope of things. The problem is, you don't have anything to compare them with, 'cause how high is up, and how big is big?

—DAVE SCOTT

That's the other thing about the moon, you have no feeling of depth perception. Because you have no telephone poles or anything to relate to.

—JOHN YOUNG

With Hadley Delta mountain as a backdrop, near the rim of Hadley Rille, Dave Scott works at the rover during Apollo 15's third moonwalk.

You know, you're not heavy, and you have the feeling that you're floating a lot of the time. . . . You're just really kind of touching your feet every so often. Moving over the surface was fairly effortless, except for the cumbersomeness of the suit. We weren't *heavy* at all.

—ED MITCHELL

Up there, with the suit on . . . you just couldn't take that big a leap. Not that you weren't physically capable of doing it at those weights, but you were going to wind up on your head or your back or your side or something.

—PETE CONRAD

The sensation of slow motion . . . when you move, the combination of the restriction of the suit and the gravity, you just *wait* to be brought down to the surface . . . you've got a lot of time, which means that time slowed down, which gives you this sensation of slow motion. Time has slowed down for you, and you're getting to observe a lot more, and you're aware that you're still not down yet.

—BUZZ ALDRIN

You're airborne for a long time. In fact, that's why you can run so good. It's because you can physically feel—I could physically feel when I jumped up, my legs resting in flight. . . . I could feel the muscle go, and then in flight, I would feel them rest. Almost like I rested more than I put force in them. And then I would land again, put force in it, and it would rest. It was like you would never get tired.

—ALAN BEAN

I realized right away that I wasn't really that good on my balance, so maybe I'd ought not run too fast. So I ended up either skipping or hopping a lot—what I call the duck waddle. Just a little amble. John's balance was outstanding. He could jump up and do a split, go down and come back up.

—CHARLIE DUKE

The craters are hard to see. They look great on the map, but they don't look worth a damn when you're running along next to them. You can't judge distance, and you can't tell how far you've run, because you've never run on the moon. So not only can't you guess the distance, if you've been running for fifteen seconds, you don't know if you've covered fifty yards or fifty feet. . . .

—ALAN BEAN

Standing on the moon in one-sixth G, you could start to fall over and you wouldn't feel that. You've got to sense that in your [inner] ear in one G, and the trouble with one-sixth G is, it's so low, it's almost below the threshold of you sensing it, until something starts to happen. . . . If you closed your eyes and somebody pushed you very lightly so that you started to fall over, you wouldn't sense that. . . . You really don't sense movement that much in your head, like you do in one G. I mean, obviously there's a gravity field there. If you fall over, you're going to wind up lying on the ground. But it's not like falling over in one G. It's like falling over in one-sixth G (laughs).

—PETE CONRAD

Running around is a lot of fun, but you're usually watching exactly where you land, because you're going to land in a hole one time, and the next time you're going to land on a little mound, and you don't want to land on that rock—you'll twist your ankle. So you're constantly not looking up as much as you're watching where your feet hit.

—ALAN BEAN

I think in a physiology sense, humans adapt to that lunar gravity quite well. And in that sense it feels natural very quickly—the ability to move and locomote. In the environment sense, it's less natural. It's very different than here. But I have no doubt that humans would adapt to it as a home in some reasonably short period of time too.

—NEIL ARMSTRONG

Running in one-sixth gravity are (right) *Buzz Aldrin and* (opposite) *Ed Mitchell.*

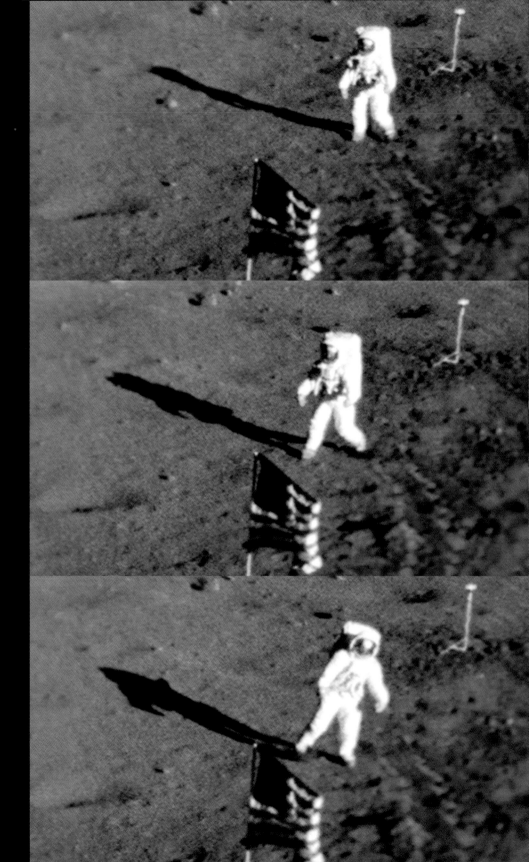

When people talk about long-duration operations on the moon, the thing they better worry about is the dust. When you're talking about designing for a stay up there for a long time, it's going to be a continual problem. It's gonna cost you seal leakage and stuff like that. I mean, you gotta get the dust out of there. It's so fine, and it's so—I mean, it's not like beach sand. It's not rounded particles. It's all impact particles, and they all have sharp corners, and if you put something on it, it never leaves. So dust is gonna be a continual problem to equipment operating on the surface of the moon, and to people going in and out of pressurized places, and to any machine that has to stay out there and rotate.

—JOHN YOUNG

Every movement of the suit was work. Just opening the hand to grip something was a struggle.

—JIM IRWIN

There are a lot of difficult things wearing a pressure suit. . . . Mainly you have to learn how to work your hand. In a suit that's pressurized, your hands get awfully tired; by the end of the EVAs [moon walks], you can barely move your fingers. So you have to learn how to work with your hands when your hands get so tired they don't work any more.

—JOHN YOUNG

[Before the flight] we both wanted our fingers, you know, right at the end of the glove, for maximum dexterity. . . . The pressure of the glove against the end of the nails caused them to become very painful.

—JIM IRWIN

See how *dirty* the suit is? That was one of our major problems, was containing the dirt, once we got back in the LM. It was hard work.

—DAVE SCOTT

It gets very sensitive after a while—like the Chinese water torture. Constant pressure. You really, in those situations, you disregard that, because you've got so many things to do.

—DAVE SCOTT

I don't think anybody realized what the dust would do up there. If you look at the movies of me walking around out there, getting the contingency sample, there's times when you can really see, I put my foot down and you can just see this great, gray cloud go out from it. But it's low . . . it doesn't billow up, because there's no air. But you could see this cloud going with you; every time you put your foot down, this cloud would go out. . . . So we got dirty. I mean, we really got dirty. You can see how dirty we are in the pictures.

It was quiet in the suit. But again, you have a fan running, moving air, and you have pumps running, moving water. . . . But it was nice

You know, you're out there setting up the ALSEP [Apollo Lunar Surface Experiments Package]. You feel good when it's complete. And you think to yourself, You know, I really aligned those things really good. All the little experiments are level, and I've got them pointed in just the right direction, the shadow things are there. Look at them, I didn't knock any of them over. And here I am three minutes ahead of schedule. Okay, now *that* meant a lot to me, but if you tell people that, it doesn't mean a damn thing to them. You know? They say, "So what? That's what he's supposed to do, is set these little experiments." They think you're doing something wonderful, and really you're just trying to make these things balance on their little legs *(laughs)* and not fall over in the dirt. To them, that's not moon exploration; that's something you do *(laughs)* between moon explorations. . . . That *is* the moon exploration. You're trying to plant these little experiments, which, at that time, you don't care whether they're seismometers, magnetometers, solar winds; you just have to put that particular device level and pointed north or something like that. And so you're really not doing anything so—more than, like, housekeeping, almost.

—ALAN BEAN

The overriding thing was always the operational considerations. Getting the job done. . . . Where are you, what's the next photograph that needs to be taken, where do we take the next sample, where do we drive a core tube in the ground, have we adequately covered the area, are we missing something? Those were the things that were going through your mind. . . . Although there was a lot of, as you say, adrenalin pumping and a lot of excitement, it was excitement toward a very specific purpose: We're there to do a job. That idea is never out of your head for an instant. Very much aware that this is a professional exploratory mission, and you're explorers, and you've got a hell of a lot of people and a hell of a lot of money tied up behind you, trying to find out what this new planet's all about. And you're their eyes and ears. You're there to observe and report.

You're very, very cognizant that you've got a batch of scientists listening to every word, looking at every frame number of film. You're very cognizant that you've got a couple of billion people listening in. . . . It's not *on your mind*, but it's always in the background. You know what setting you're in. So you're conscious of the PR, a bit, you're conscious of the fact that you're not going to say anything that makes a fool of yourself, or gets you in trouble with NASA, or causes the country diplomatic problems. You know, you're not thinking about that consciously but, you know, you're aware.

—ED MITCHELL

One of Alan Bean's photographs documenting the Apollo Lunar Surface Experiments Package he and Pete Conrad set up during Apollo 12's first moonwalk.

It was rocking and rolling, like being in the sea. . . . Same kind of roll, pitch, heave, and yaw sensation. Much sharper and distinct, though. The sea is fluid; the moon is hard.

—DAVE SCOTT

What really seemed fast was the fact that you kept bouncing off the surface—you didn't want to go any faster! [The rover] left the surface at every bump.

—JACK SCHMITT

The [rock] fields, especially going south, were a surprise to me. . . . Survey Ridge was very troublesome. . . . I was afraid we were gonna go through there and knock an axle off. Because you couldn't really see too good when you're driving the rover.

—JOHN YOUNG

We were always running the car at full throttle, trying to make the most of our time on the moon. And what was out ahead of us was a little uncertain. Many times we'd come up over a hill and there'd be a crater, and Dave would have to throw the controls hard in one direction, and we'd go up on two wheels. And it felt like, you know, we might actually roll over. . . . The rover really seemed to be another spacecraft, even though we were operating on the surface of the moon. Every time we'd hit a rock or a bump, we'd just fly into space. So I estimate we were floating through space a good bit of the time.

—JIM IRWIN

Yeah, it was harder driving the third day. . . . 'Cause you couldn't see. You couldn't see as well. You get a white-out. Almost like skiing. 'Cause you don't have all those holes in front of you. . . . Well, you hit more bumps (laughs). Poor Jim. Tolerant guy. Tolerant guy.

—DAVE SCOTT

When the fender came off the rover, it was just raining dust down on us.

—CHARLIE DUKE

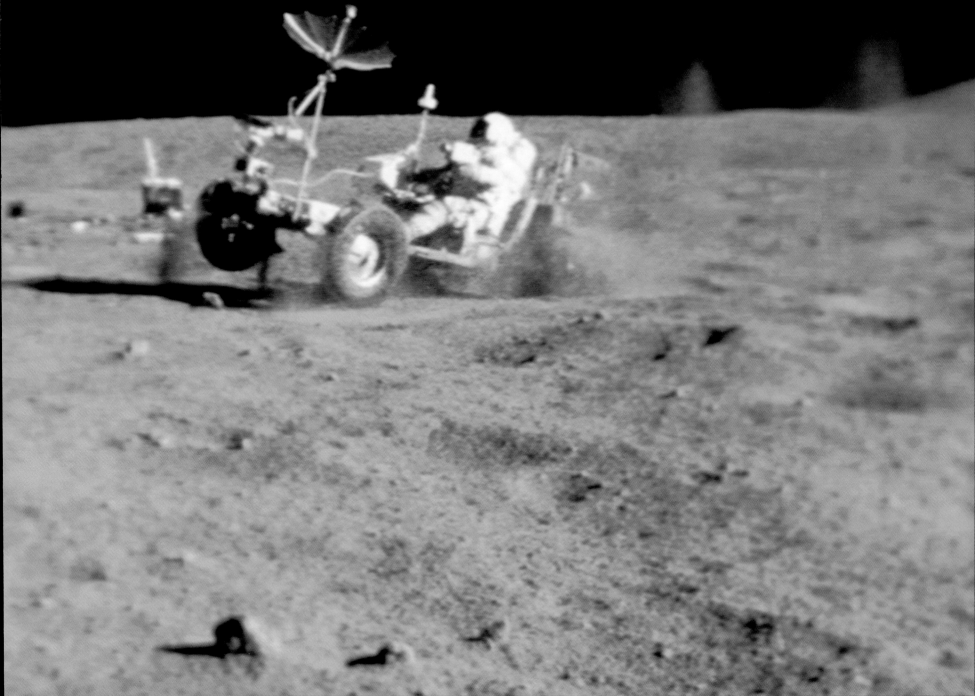

To the casual observer, rocks are rocks. But we'd spent enough time in the field, and [geologist] Lee Silver had had us get suites of rocks and tell the difference between all of these. . . . Well, they all looked the same to me, but after a while they don't look the same. . . . So you get to the moon, and you look around, and they're all different. Because you're tuned, and you got a second-, third-, fourth-order differentiation in your mind, because you're looking at shape, size, texture, granularity, and all those things that you've become familiar with. . . .

[It] became accepted. . . . "Of course it's four billion years old." . . . Unless you stop and think about it, . . . you [can't] accept it as being that old. . . . For people who don't spend the time on the geology, they don't learn to accept the fact that it's that old. I mean, I remember when we first started, somebody said, "A billion years," and I thought, Huh? Nothing's that old. You know, I'd never studied geology. So you come to accept the fact that this is three to four billion years old. I don't even know how far that is. What is that? Who knows. You can't comprehend that kind of time. . . .

Consciously, I was talking to them [the geologists on the ground]. I wasn't reciting anything for the world. . . . That's what my training was, to talk to them. . . . So when I spoke about [geologic] things, I was directing my comments, specifically, to Lee Silver, Gordon Swann, Joe Allen, Jim Head, the guys we'd been working with. And I didn't really care about the rest of the world. I wasn't even conscious the rest of the world was listening. . . .

One of the things I guess I said to Jim is, "Hey, we got what we came for." Absolute, spontaneous comment. . . . 'Cause I was real happy . . . boy, we'd been drilled on that and drilled on that and drilled. Early crust of the moon? Get a piece of that? And prove it? Oh, *yeah* . . . once you picked it up and looked at it, absolutely no question what it was. . . . It was so different from everything else. And we'd been primed. That's stored back there [in your head]. You're not consciously looking for it, but when you pick it up, if you hit that chord—*ding!*—there it is.

—DAVE SCOTT

I never remember the names of anything. Those guys wanted us to go out and say, "We got a microbreccia pyroxene crystalline structure," and I used to say "stuff." I wasn't noted for my geologic ability. In the back of my cuff checklist I had [geologist-astronaut] Jack Schmitt put a list of names of either geologic phenomena or rocks that would have really lit the scientific community's candle. . . . And I got busy enough not to carry that joke off, either. But I had the checklist with all the names written in the back of it. I had sort of a special secret page, so I would not only just refer to it [rocks and dust] as "material," but I was going to pull some really long-winded geological terms out of the basket, and wax on like I had my PhD in this stuff, because, like I said, I was not known for being the world's best geology student.

—PETE CONRAD

It's hard to whack on the moon. Because the [hammer] only weighs one sixth. It's got the mass, but you can't get the force behind it. You're not standing on the ground. If I'm planted on the ground, when I hit something, I'm transferring that impact to the ground. Right? Same mass, same momentum. But it just didn't feel like you could whack it as hard as you can on the ground. Maybe with the suit or whatever. It was tough breaking things off.

—DAVE SCOTT

Dave Scott prospecting for samples at Hadley Rille.

At the moment you're in a living adventure, a dynamic adventure that requires your attention to detail, technical aspects, and while your mind is attuned to the technical deal of the stop at . . . North Ray Crater—What are we supposed to accomplish here? What are the objectives?—while all of that's going on in your mind, you're also—Yahoo! Here we are! you know, and, Man, North Ray is even a lot more beautiful and mysterious than we expected. And those surprises that would come to you . . . the House Rock. . . , those were like little goodies, little plums that were given to you, like going into a candy store and getting a big assortment when you were just expecting a chocolate drop. . . . Over the next ridge was going to be a special little find. . . . In the back of my mind was, you never knew what was going to be behind the next rock. So it was like an Easter-egg hunt—What am I going to find next?

—CHARLIE DUKE

The true Genesis Rock of the Apollo program was the one that I found . . . the one that was 4.6 billion years old. That was over at the base of South Massif. The question is whether somebody [other than a professional geologist] would have noticed. . . . It was a distinct texture and a distinct color. Again, I can just go back to the question of how . . . experience enables any professional to filter out the millions of extraneous bits of information and see in microseconds what is significant.

—JACK SCHMITT

It doesn't make any difference whether you're in a suit or whether you're in shirtsleeves. You're still going through the same mental process. It's the *mind* that you're taking, not the hands. . . . Your mind is not in a spacesuit.

—JACK SCHMITT

In general, we had agreed that we had limited time, and we had a lot of things to do, and we were going to stay in the close vicinity of the LM. And I don't know what the rules were, but it seemed to me that that crater was a pretty deep crater, and there was a chance of bedrock showing, a chance of outcrops. And it would be very worthwhile to get a picture of the inside walls of that, something the geologists—and it wouldn't take very long to get back there and get it. So I did it. It wasn't very far away. And it didn't take me very long, so I thought it was worthwhile. I don't know if it was breaking a rule or not (laughs).

—NEIL ARMSTRONG

Having run a distance of about 200 feet east of the lunar module, Neil Armstrong documents the landscape at the rim of an 80-foot-diameter crater.

Let me give you another analogy. You live in Iowa all your life. And you drive out to Arizona. And you drive up through those trees to the Grand Canyon—blindfolded, okay? And somebody takes you right over to the rim, and then takes the blindfold off, and you look down at the Grand Canyon. That'll blow your mind, won't it? . . . Well, we drive . . . up toward St. George [crater], on the first day. . . . You don't have a big peripheral vision. Stop the rover, and get off, and turn around and look, and the goddamn Grand Canyon— Hadley Rille! I mean, that's an absolute mind-blower. Even though you know it's there, but you can't see it, 'cause you're driving this little rover next to the ground. Hadley Rille's over there—you can't *see* Hadley Rille. You can't even see craters. All of a sudden you get off and you turn around, and there it is! In all its glory. The Grand Canyon of the moon! That's mind-boggling! I mean, that'll blow you out. So I would draw the analogy with the poor little guy from Iowa.

—DAVE SCOTT

A mile wide and a quarter mile deep, Hadley Rille looms beyond Dave Scott and the rover during the first Apollo 15 moonwalk.

We found it difficult to walk upslope at that angle. The material was loosely compacted, and it was like walking up a sand dune; it would slide away from you. . . . It was much easier to go up the hill in the automobile than it was to walk. . . . You know, the treads would just sink in maybe half an inch, whereas if we'd step on it, we'd sink in four or five inches. I don't know why; I haven't figured out why that was so.

—JIM IRWIN

At one point, the rover started sliding down—yeah. It was concerning. But Jim went over and held the rover. In fact, there's a picture.

—DAVE SCOTT

Below: *Dave Scott's photo shows Jim Irwin holding the rover to keep it from sliding down the steep slope of Hadley Delta mountain.* Opposite: *John Young* (pictured) *and Charlie Duke had this view from the side of Stone Mountain, hundreds of feet above the Cayley plains.*

I took a telephoto from there, and it's my favorite picture . . . and I got the LM in it. And with a 500 mm lens you can barely see it. And the impression you get from being up there and looking back is that, Boy, you're a long way from home. 'Cause that [the LM] is home. And if you look out at the whole expanse of the plain, there's nothing there . . . but that one teeny little thing. . . . An enormous scene, which is very clear, very bright, in which . . . there's one teeny little thing that's called a lunar module, all by itself. . . . And boy, that's impressive . . . There's a manmade thing, and then the rest is pure nature. Absolute pure nature. And the manmade thing is swallowed. Behind it is Pluton, the big crater, and Pluton is further away, but it's *enormous* compared to that little LM. . . . And the lunar module's big—that's a big bird . . . And yet, in that scene, it's [a] teeny-weeny, small little thing. Hardly discernable . . . I mean, it sort of gives you a perspective of your own dimension in things, which is very small.

What I've always wanted to do was to get the regular picture and put that on the wall, and then draw a square around where the telephoto took, and then put [the telephoto] on the wall. . . . And in the [telephoto], you still see the little LM, 'cause you can't see it in the regular picture. Anyway, this is one of my favorite pictures.

—DAVE SCOTT

Standing on the side of Hadley Delta mountain, some 300 feet up, Dave Scott and Jim Irwin had this view of the undulating, crater-pocked plains. The white square indicates the field of view of Scott's telephoto picture (opposite), *which captures the lunar module* Falcon, *some three miles away, and, beyond it, the half-mile-wide crater Pluton.*

The one thing that was a surprise was, we couldn't navigate like we thought we could, because the dunes were much higher. And you just couldn't see where you were. You'd lose the LM when you went down into a depression, and trying to navigate up to Cone crater—we had very precise points laid out to try to get to, and we never were sure whether we were near one of them or not . . . within ten yards of it or exactly on top of it. Or whether it was over in the next valley. . . .

On foot and almost a mile from the lunar module Antares, *Ed Mitchell collects samples, just as he and Alan Shepard are forced to abandon their hunt for the rim of Cone crater.*

about . . . knowing your position within five or ten meters, and we never really did. . . . They wanted us to turn back, and we kept trying to go on, and go on, and go on. . . . But we finally had to turn back, and there we were, about fifty or sixty feet from the edge of the crater, but didn't know it. . . . It was right there in front of us. . . .

But to us, looking at it, it just looked like another sandy hummock over there, and it fooled us for an hour. . . . And we had to

yards we thought, the next one is the edge of Cone crater, and it was just another ridge. We assumed that the next one would be just another ridge also, and it turned out, no, in this case, the next one really *was* the edge of Cone crater.

—ED MITCHELL

The thing is—it's hard for you guys to understand—that clock was ticking, ticking, ticking, and you were very, very conscious of the fact that you were operating against time, continuously, and we had deliberately put into the timeline about fifteen percent more than we could expect to accomplish. But our egos demanded that we try to accomplish it . . . which meant that you never stopped. Always moving.

—ED MITCHELL

I was surprised that time went so fast. We never had enough time. And, boy, we had trained hard to make sure . . . we were efficient. And when we got there, we never had enough time to explore a site like we would have liked to. Because it was so exciting. There was so much there. And you just want more and more and more and more. . . .

—DAVE SCOTT

The other emotion, the feeling which I think everybody had, was you were never going to have enough time to do what you wanted to do. Dick Gordon said when we first started working [on the Apollo 15 backup crew], sometime in our training cycle, "You'll find that time is relentless." And it is. It just never stops. You keep running out of it. . . . Time is not relentless in an ordinary field experience. . . . But in most field experiences, you really do have the option of going back.

—JACK SCHMITT

Documenting his sampling site during Apollo 17's second moonwalk, Jack Schmitt also photographed two of his geology tools—a long-handled scoop and a rock collection bag—lying in the foreground. In the distance, Gene Cernan works at the rover.

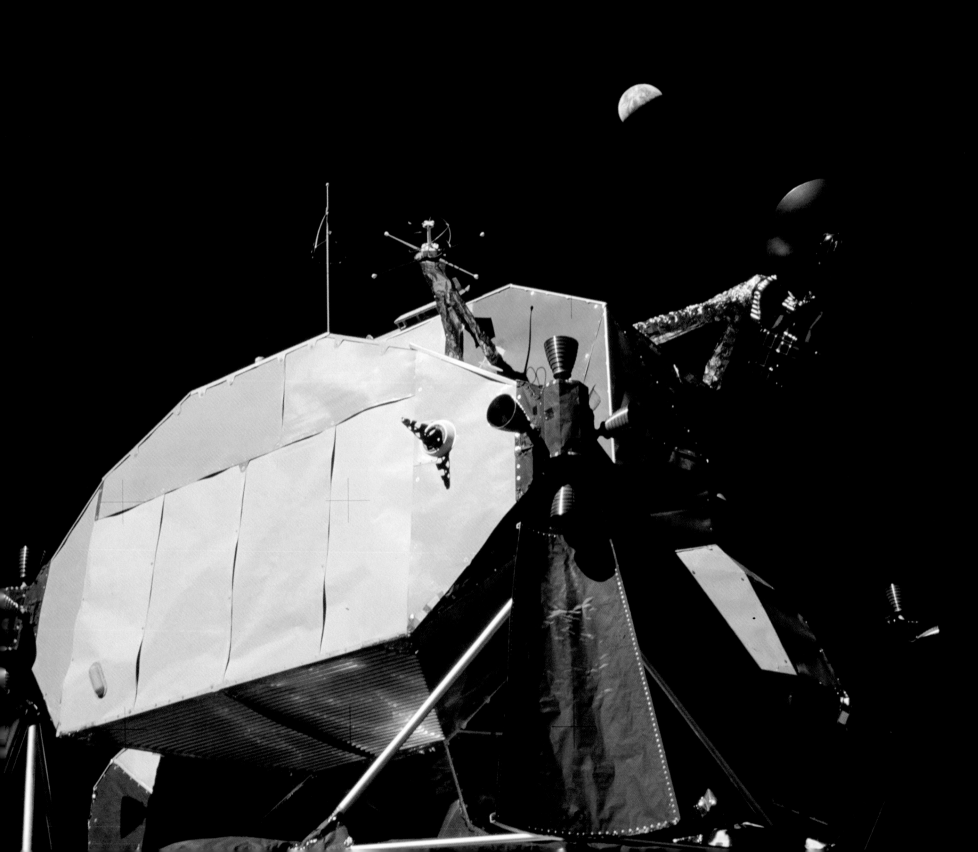

Sure, it's breathtaking to look out and see the Earth. And probably, if somebody'd said before the flight, Are you going to get carried away by the Earth view? I'd have said, Nah. That sort of stuff—Are you going to get carried away looking at the Earth from the moon? I would say, No, no way. But yet when I first looked back at the Earth, standing on the surface of the moon, I cried. And if everybody had ever told me I was going to do that, I'd have said, "No, you're out of your mind." . . . Whether it was relief, or whether it was the beauty of the Earth, the majesty of the moment—I don't know, just every— you know, I never would've said I was going to do that. But I did.

—ALAN SHEPARD

To look directly overhead was almost impossible, because of the backpack. . . . I saw the Earth for the first time when I came down [to take my first steps] and grabbed the ladder to get my balance, when I lost my balance. I reached up to find the ladder, and went around the back of the ladder, and in doing so, I looked up and saw the Earth for the first time. Just a spectacular view of the Earth.

—JIM IRWIN

When you're standing on the surface, you see the mountains, and the Earth is on top of it—it's like that picture I got of Jack, where we got the mountain, the flag, the Earth, and Jack all in the same picture. That's a new dimension, because then you can relate it to where you are. . . . It just drives home: That's the Earth, and I'm here.

The Earth looks from the moon like the moon looks from the Earth, and vice versa. You can't really tell the difference at that distance, even though the Earth is a lot bigger. They all look very small at 240,000 miles out. When people go to Mars, and they start looking at two little dots out there after a couple of days, that ought to be a very interesting trip.

—JOHN YOUNG

I can remember running along and looking up at the Earth. That's when it would really hit me that we were on the moon. And I would say, This is the moon; that's the Earth. I'm really here. And then I would say, I've got to quit doing this . . . because when I'm doing this, I'm not looking for rocks.

—ALAN BEAN

Looking back at the Earth, it is your identity with reality. It's home. It's where family and love and life really is, viewed from a vantage point a quarter of a million miles out in space where reality itself is almost a dream, . . . a dream in which you are a very vulnerable character. You're a very vulnerable ingredient to that dream. . . . You are really—I was really on the moon. But even that reality, as you look back home, is almost like a dream.

I wonder what it would have been like to walk on the moon and not have the Earth in the sky.

—GENE CERNAN

Earth hangs above the lunar module Challenger *near the end of the third and last*

[Before the flight] that was a long, laborious decision process—do we take off the suits, not take off the suits? We . . . just took off helmets and gloves. It did make sleeping and everything difficult, but it took *so long* to get *in* and *out* of the suits that we just didn't want to waste the time. . . . And, you know, you run the risk of damage, or a zipper not working, or something like that. So it just seemed a lower-risk course to leave the damn things on. I think one of the reasons they removed them on later missions was because all of us complained that it was terribly difficult to sleep with the suits on. . . .

We were very concerned about the angle that the LM was on. . . . Neither one of us slept very well that night. And we kept [hearing] creaks and groans of the spacecraft. . . . I think there were probably even a few meteor impacts on it, ah—pings here and there that we didn't know what they were. Probably that's what it was. . . . And we got up and looked out the window three or four times during the night, to make sure we were still upright and hadn't turned over—nothing was going on.

—ED MITCHELL

The other guys didn't sleep—I couldn't sleep in the suit either. It's tough. . . . The other guys left their suits on; we took our suits off. . . . Getting out of the suit, and then going through your normal cycle. Your *normal mental cycle.* . . . It was comfortable. Ooh, it was really comfortable, as a matter of fact. One of the things Jim and I both noted—we practiced in the simulator on the Earth sleeping in the hammocks. And it was okay, but hammocks on the Earth, supporting all that weight, well, you're okay—no big deal. Hard to turn over; you know how hammocks are. Gosh, got on the moon, and it was so nice because you weighed one sixth, and the hammocks didn't sag as much. It was like a feather bed. It was *really* comfortable. . . . Boy, I got in that little dude, and bang, away I went.

—DAVE SCOTT

I felt like I was at home. Even though I knew that I was on the moon, I could almost convince myself that I was back on Earth, back in the simulator, where I went to sleep many nights, simulating the same conditions that would exist on the moon. . . . That's why I slept so well the first night. The next night I was too exhausted to sleep well.

—JIM IRWIN

You've got to understand—I'm sure you do—going from a day and a half to three days is much longer than twice as long. Because you have many more things to contend with. I mean, it's a totally different mission when you go from a day and a half to three days. . . . Because I don't have to sleep during a day-and-a-half period. I can go two days. Can't go three days. No way. You've got to sleep. Or you won't perform. So, one of the things we had to do was learn to sleep and set up to sleep. At the Cape, we spent a night in the simulator to go through that procedure. And we tried to pretend like it was home time. And we took our suits off, which other people hadn't done, got in our skivvies, got in the hammock, turned the lights out. . . .

So [on the moon] we'd have dinner, get out of our suits, clean up, brush our teeth, go to bed, and turn the lights out. Go to sleep. And that way, we lived like you normally do, as closely as possible to your normal living cycle. Because if you don't do that, you won't perform the next day. . . . You may think you're performing, but you're not performing. . . . So in that sense we *had* to learn to live on the moon. The other guys prior to our flight didn't have to live on the moon—they *stayed* on the moon. They stayed on the moon for a day and a half.

—DAVE SCOTT

Grimy from three days of exploration, Gene Cernan and Jack Schmitt's space suits and helmets are piled in the cabin of the lunar module Challenger *after Apollo 17's final moonwalk.*

Just think about going to sleep on the moon inside of an oil can. And that's what we did. An oil can that buckled every time you pressurized it: *boom-boom*. And you're inside that oil can. You're on the surface of the moon, and it's time to rest or go to sleep, which is the biggest waste of time in the world—who wants to go to the moon to sleep? That's when you have a chance to *think* about these things. Think about what you just *saw* and where you *are*. Yeah, you think about those things. You don't think about them while you're picking up rocks and doing all these other things that you've got to do while you're there. But you think about them when you have a chance to stop.

—GENE CERNAN

Having left Apollo's last footprints on the moon, Gene Cernan takes a breather inside Challenger.

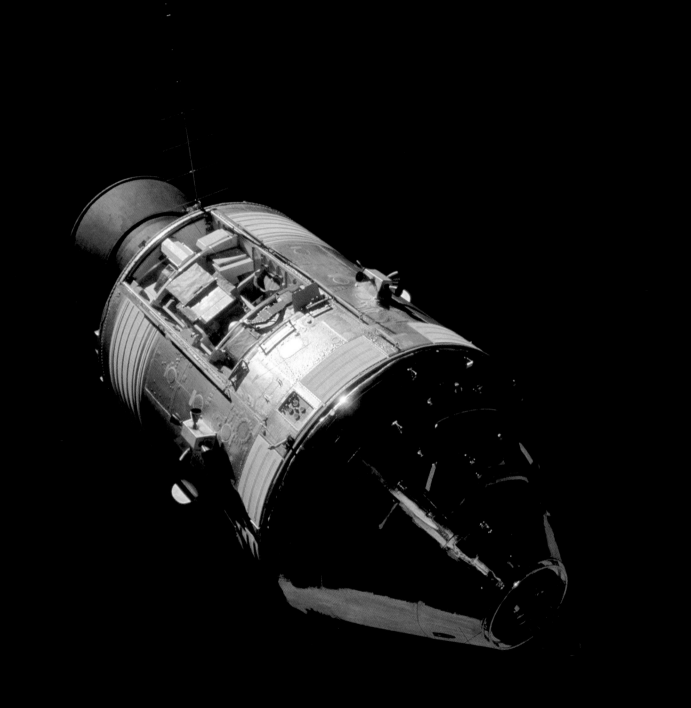

7

SOLO

The most exhilarating thing in the world. God. To be there, by yourself, totally responsible for this thing. Dead quiet. And this spectacular, unreal world. Nothing could be more exhilarating.

—KEN MATTINGLY

Everybody that flew in the command module hoped that they would make some contribution in some way besides being a service organization. Otherwise, you're a truck driver. You brought the big boys to the moon, dropped them off. You didn't screw anything up; you were still here when they got back. But you would like to say, I made a contribution. I did something. So you grasp at straws. . . . I find it a fascinating subject. But did I have these great desires to learn about the moon? Nah. I'm not a scientist!

[But] I got this extraordinary opportunity. I'd like to do something for it. Sure, everybody had aspirations about how we were going to go do something different. See something that nobody's ever seen before. . . . And I can identify with that, and certainly my experience is that no picture has ever captured what I've seen in its vividness. But on the other hand, you can't take any of these emotions that I have and make dimensional analyses on them. . . . A lot of what I did was just because I admired Farouk [El-Baz, geologist] so much that, you know—If you want to do this, hey, fine with me. Why not. Let's go do it.

—KEN MATTINGLY

I think this solo aspect of being in that spacecraft, as I've said before, designed for three people, by yourself, 250,000 miles from home, you feel fairly lonely. That's the closest time in the whole flight that you break out of your test-pilot, give-'em-hell, better-pilot-than-you attitude, is that feeling. . . . A lot of mixed emotions. None of which is fear. I guess, some wonderment. Awe. I suppose, thankfulness that you're there, that you have the opportunity to experience that. Some loneliness.

—STU ROOSA

I don't think—since I haven't done it, I can't say. But I can't imagine bouncing across the surface of the moon being as personally exhilarating as being solo in a spacecraft on the back side of the moon. I just don't think there are very many things you can do that are more satisfying. . . . It's just—there's nobody here but me! And there's no noise, except the little electronic fans going *poof*. And you turn some music on, and you watch this panorama go by, and it's absolutely mind-boggling. You can't imagine it; there's nothing like it in the world. . . . I mean, it's just one sight after another that is just absolutely extraordinary. And you're all by yourself! Just you!

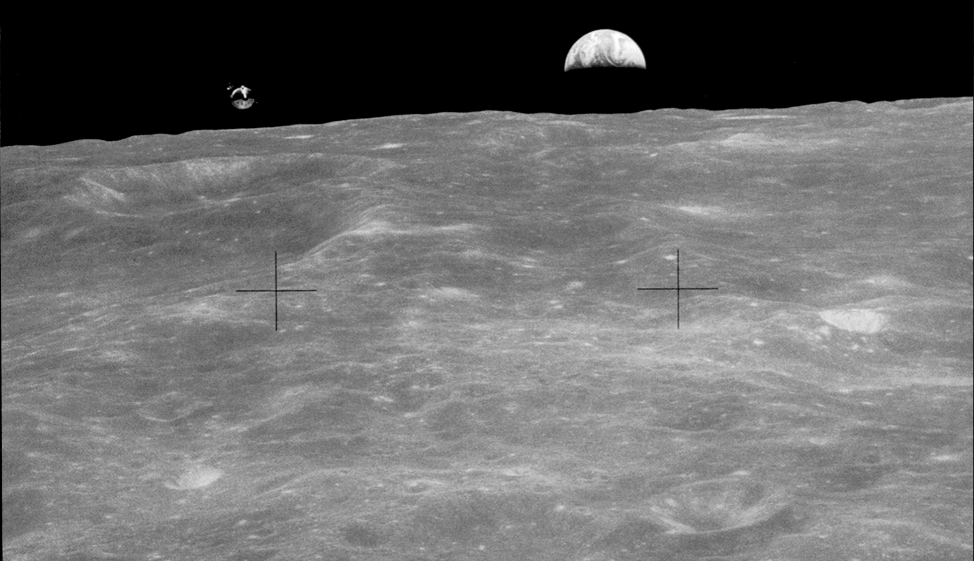

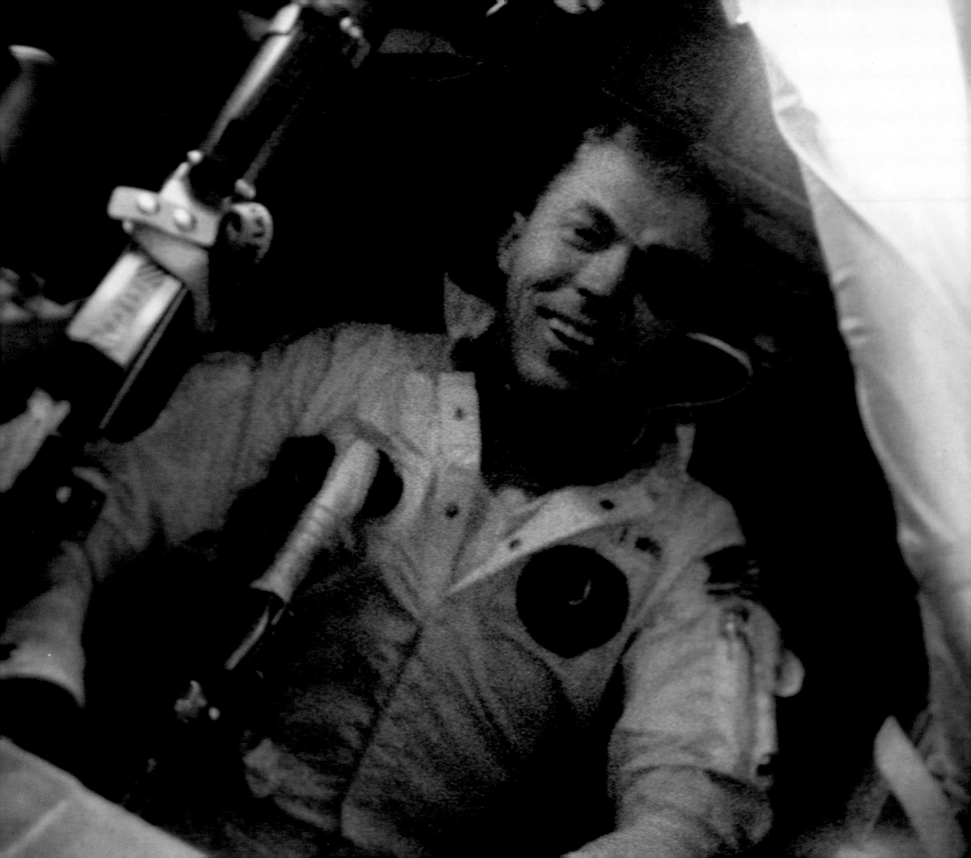

I knew that the navigation was superb enough that I knew what I was looking at through the telescope. . . . I knew when I looked through the sextant that it would be looking at *the* Surveyor crater [location of the Surveyor 3 probe visited by Gordon's crewmates]. And the things that I saw there were interpretive, as opposed to a strong visual. I never saw the [lunar module], but I saw a bright spot and what I considered a fairly long shadow. And I said, "Okay, that's got to be the LM. . . ."

You know, Mike [Collins] didn't have, or neither did the ground, any idea on God's green Earth where Apollo 11 was. Where they landed, or anything else. . . . Mike and I had a conversation about whether or not he thought they could see it. He never saw it. Probably didn't know where the hell to look. But our navigation [on Apollo 12] was precise enough . . . I knew the damn crater I was looking at was the Surveyor crater. There was never any doubt in my mind.

—DICK GORDON

The next time around I tracked the lunar module. And I mean, there was a lot of questions about whether you were going to be able to see [it]. But I really didn't have any problems. It was not vivid; of course, you know, it's a little reflected sunlight, is what you're looking at. It was different enough from the lunar surface that I could see it. . . . You saw the reflective bit . . . but the real clincher was the shadow.

Oh, and later on . . . I think I tracked the LM again on the pass just before the ascent. Because I came by the front and I tracked the lunar module again. And by then it was a piece of cake. . . . The shadow was less, but it was brighter. But also—and I said this in the debriefing, and I don't think anybody believes me—but I saw the ALSEP [experiments] reflecting light. . . . I, in fact, drew a picture [at the time] of where the LM was and where I said the ALSEP was. So then that convinced them.

The rest period wasn't that long. And I didn't sleep well anyway. I had to lie about how much sleep I got because everybody would've been worried. . . . Your time in lunar orbit is so incredibly busy. . . . And when you perturb the timeline, it's just hell to get back on schedule. . . . I didn't have time to eat. I mean, I had time to eat in the flight plan, and I didn't. I never fixed one of the freeze-dried meals, where you had to pump the water in, and so forth. [We carried] some of these spreads, like ham salad, and tuna salad, and . . . you could smear it on a piece of bread, and eat it, and be done with it. I didn't eat very much at all.

—STU ROOSA

By the time I got back, I was so goddanged tired I could hardly walk. I know I couldn't possibly have had more than a dozen hours sleep, cumulative. And it's not concern *(laughs)*. It's just, you don't want to *miss* anything! You just can't get that tired. You just—you finally get to the place where you say, If I don't get some sleep, I'm going to do something really stupid! But, jeez, this may not come back again any time soon! I don't want to pass this up! Well, peek out the window. They won't know if I'm awake.

—KEN MATTINGLY

Wearing the eye patch he used while making navigation sightings, Stu Roosa floats in

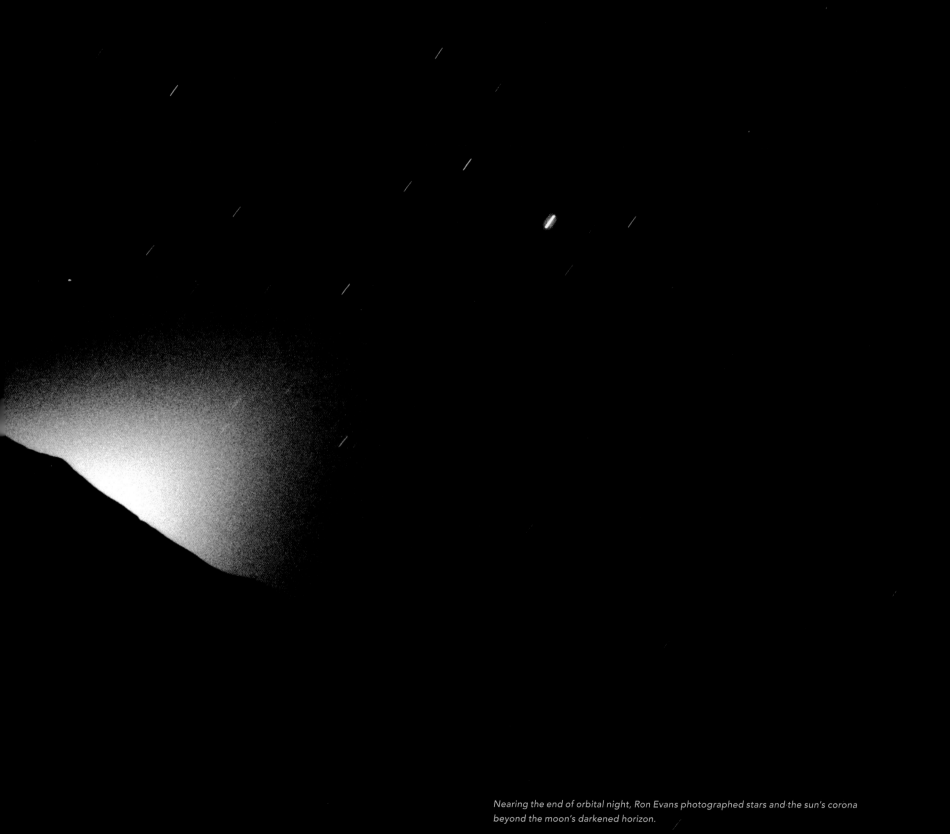

Nearing the end of orbital night, Ron Evans photographed stars and the sun's corona beyond the moon's darkened horizon.

by myself, in lunar orbit, the pie-shaped part of the moon that was in total darkness. Turn all the lights off in the cabin. Look out through the window at the rest of the universe and realize that there's a lot more there than you had ever imagined before, because you can see so many more stars. . . . There are so many stars out there that you lose track of the constellations that you navigate by, because they all get washed out. . . .

That is isolation. You could only tell where the lunar horizon was by the starlight that it cut off. Otherwise you couldn't see anything. . . . Total isolation. I thought it was great. I loved it back there. I guess that's the mentality of a fighter pilot—likes to be in the machine by himself, likes to fly by himself, doesn't particularly care to have to relate to anybody outside the machine. I was there by myself after Dave and Jim went down to the surface, which was good, and on the back side of the moon, where I wasn't even talking to the ground, it was better yet. I still did everything that I needed to do, but I sensed a freedom about what I did, that you don't get if you've got the radio on. . . .

I've never believed we're the center of the universe. It's always been in the back of my head that planet Earth is just one of billions. And, you know, we're making some first steps. I feel even stronger now than [I did] then. I think there's no question about it. . . . So all of a sudden you're thinking, you know, we can tell ourselves intellectually there are so many billions of stars, and out of those there's x percentage of medium-size stars with a planetary system around it. You can go through all that hocus-pocus you want to. But until you get out there and you see the star field, and you understand that there is so much more out there—It's very easy to believe that there might not be any other intelligence out there. But you know without a doubt after you've seen it, that, I don't care what probabilities you give me, if there is any probability at all of life out there, there's gonna be lots of it out there. Because, I mean, this thing goes on forever. . . .

I basically had that feeling before the flight, but since the flight, the more I think about it, the more I realize that there's just absolutely no other way it can be. . . . It's come about more by reflection afterwards. . . . You gotta think about something like that for a long time. That's not the kind of thing where there's a flash and all of a sudden there's understanding. That's the kind of thing that you develop over a period of time. . . . If I never had the chance to go to the moon, I might have a very different view of the universe.

—AL WORDEN

During that quarter of the moon in which you're not in sunlight or Earthlight—you're in that total darkness—just by the nature of our environmental control system, we're running at a hundred percent humidity. And so as you cross into what I call the first terminator, and you're in darkness but you've still got Earthlight illuminating, the spacecraft cools off, just a one- or two-degree drop in temperature, and then it gets moist on the inside. In fact, it gets visible condensation on the walls, so you get kind of a clammy feeling.

And then you go into that other quarter, where it is so abjectly dark. And it's clammy. And you *feel* that. You feel that darkness. I don't know how you feel darkness, but you do. . . . The darkness is a very penetrating feeling. And then a remarkable thing happens. You come around, and instantaneously it's sunlight. . . . Suddenly the sun is there. I mean, without any warning. One instant it is total blackness; the next instant the sun is there. Just (snaps fingers) *immediately*. That fast. And that's always a marvelous experience. You feel good. You think in your mind, *You know, we're really not creatures of darkness.* You immediately start feeling better, just because there's sunlight coming through the windows. And this repeats itself every two hours.

—STU ROOSA

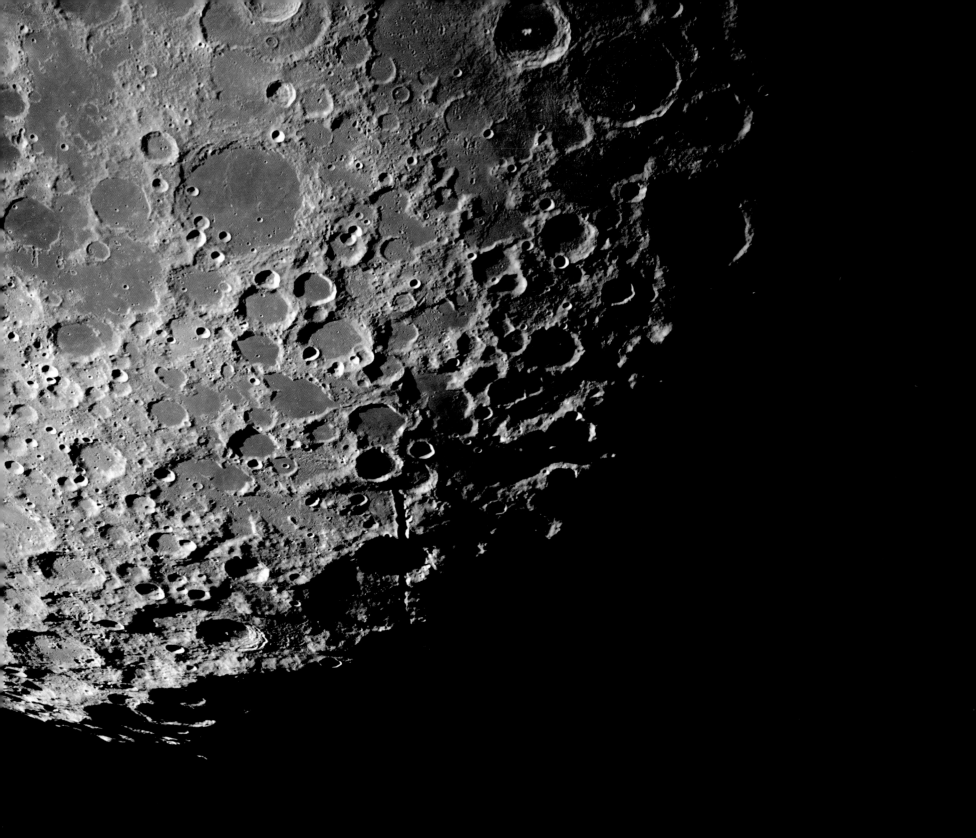

8

HOMEWARD

I didn't want to leave. . . . Real sad to think of having to leave at that point, knowing you'd never get back.

—ED MITCHELL

We were two hours ahead of schedule. . . . There was a place [in the ascent checklist] where we had to stop and let time go by before we picked it up. So we had two hours to kill. . . . Needless to say, having gotten to that point, my ass was dragging. And I remember . . . sitting down on the floor over on my side, and figuring, Take a break. . . . So I'm just sitting on the floor, cooling it. And Bean's over on his side, standing up. And he's fidgeting with the checklist, and doing one thing and another. And really he's getting nervous over there. . . .

It finally dawned on me—I went back to my "eight days in the garbage can" [on Gemini 5]. My fear . . . on my first flight was that nobody had ever had retrorockets cool up there in the breeze for eight days, and I was panicked that, when we pushed the retrofire button, that those goddamn things wouldn't fire, and I figured that if they didn't, I was going to slit my wrists, 'cause I couldn't stand it any longer. . . . It came to me sitting on the floor that Beano was going through the same drill I'd gone through on Gemini 5 and gotten that out of my system. His problem was, he was worrying about the ascent engine lighting. . . .

Anyhow, Beano's got the flibberti gibbets, and he's dancing around and looking at the switches and out the window and turning around and, really—no fuzz on it. And I finally said, "Beano, are you worried about the ascent engine?" or I made some remark like that, and he said, "Yeah." And I said, "Well, there's no sense worrying about it, because if it doesn't work, we're going to become the first permanent monument to the space [program] on the moon." I'm not sure that did any good for him (laughs). . . .

I'd been through two [spaceflights] . . . and I had all the confidence in the world in that ascent engine. . . . If I worried about anything, I worried more about all the explosive bolts, and the separation [from the descent stage]. . . . I was sure the sumbitch would light (laughs). I wasn't sure we weren't gonna blow right through the bottom of the descent stage.

—PETE CONRAD

We used to joke about the fact that you have, I don't know, ten thousand guys at the Cape getting you launched? On the moon you have two.

—DAVE SCOTT

I had that concern [that the ascent engine might not fire] and had focused to some extent on it for a substantial period ahead of flight. And in fact suggested at one point that we consider replacing the electrical system for actuating the valves into the engine with a mechanical system for an emergency method. But that was considered and rejected. Nevertheless, if something—if the engine doesn't light, you're not in a dangerous position. You have a lot of time. You can talk it over with the ground, decide what kind of alternative methods you're going to try. And you're not out of options at that point. So compared with a lot of other serious things in a flight, that wasn't up at the top.

—NEIL ARMSTRONG

No, there was really not any *worry;* it's just the apprehension, the anticipation of waiting for it to go. And since we'd never experienced that liftoff, we didn't know what we were going to feel. It was a *pretty severe* shock. It staggers you, . . . it makes you sag.

—ED MITCHELL

Ascent was spectacular. We flew *right* down the rille. And that was a surprise. All our ascents had been done in the simulator, without looking out the window at anything in particular. The model maybe. But we never thought about it. You never think about what you're going to see out the window. Boy, that was really a great show. Almost like somebody had planned it for us—Okay you guys, when you take off, fly up the rille on your way out—Like you would fly an airplane up the Grand Canyon. By God, there it was.

—DAVE SCOTT

An onboard movie camera records the liftoff of the Apollo 14 lunar module, Antares.

We launch on the moon. I'm looking out the window. I see us blow these little pieces of foil . . . all around. Gold and silver and black. They're flying around just like you threw a pebble in the water: concentric rings. I'm looking out the window—you know, people think you're thinking a certain thing about lifting off the moon, you know? I could probably invent some sentences that would fit it. But what I was really thinking as I looked out there was, I hope these little pieces of foil don't land on our ALSEP [experiments] and change the thermal properties. See? That was important to me, and it's something that's on my mind. And the philosophical—launch from the moon, never to return; we were the second people on the moon;

the Ocean of Storms won't be visited for another fifty years—none of the stuff that people might *think* you're thinking—that wasn't on my mind! I was hoping the engine would fire, and it did, and the next thing I was hoping was these little pieces wouldn't land on the thing.

—ALAN BEAN

My favorite feeling about Pete . . . was the story about him letting me fly the lunar module during the rendezvous phase, you know, just fly it. . . . We could see Dick, we were getting close. We'd made the last big burn, we had two midcourse corrections to do or something like that. . . . But somewhere around in there, Pete said, "Hey Al, you're working too hard. . . . Why don't you sit back, look out the window, and enjoy the rendezvous." So, after I kinda collected my wits, you know, because he'd never done that, I said, "Well, gee, this is great," you know. "This'll be really fun." I—I felt guilty about it, 'cause he'd never done it. It was a surprise.

Then he said, "Would you like to fly this thing?" I said, "Well, we'll get off course here." He said, "Well, we'll call up our delta-v program [for measuring velocity], and then . . . you can fly it back and forth, and up and down, and then we'll zero all those things out and we'll be right back where we were." . . . I said, "Well, the ground—" He says, "Don't worry about it. We're behind the moon (laughs). They can't tell what we're doing." So, I flew it around for—I was kinda nervous about flying it around, but I, you know, flew left, and then right, up and down, 'cause I didn't want to get off course either! . . . You know, it was something we'd never tried before. We'd never done this. Although it seemed like it wouldn't hurt anything, at the same time you learn in spaceflight you're not always smart enough to figure out everything. So I did that for a few minutes, and enjoyed it, and felt it. It just felt different to do it than to be paying attention to something else and have Pete do it. . . . Then we put it back to zero and went on with the flight.

And I've often thought of that time as how a guy could be in a life-threatening situation—it's the first time he's been behind the moon; it's the first time—I mean, rendezvous with Dick. Our neck's out. We're in this little lunar module, you know, we're all dirty, ah, and here he's thoughtful enough to think that I might like to do something different. You know, I mean he [empathized] enough with me. That's why I often say he was a great leader. Even in these dangerous situations, he can come outside of himself enough to empathize with someone else, namely me, and say, This guy's on his first flight, he's working hard over there. Why don't we let him have a little fun? Maybe there's a way that he can have some fun and it won't get us into any trouble. And just the fact that it would cross his mind, and then he would figure out how he could do it, and then do it, shows what kind of a guy he was.

—ALAN BEAN

Above: *Pete Conrad and Alan Bean on their way back to rejoin Dick Gordon in lunar orbit.* Opposite: *Gene Cernan* (visible in window) *and Jack Schmitt rendezvous with Ron Evans.*

As soon as we got back into lunar orbit we had a lot of dust in the cockpit, because we brought it all back with us and hadn't swept out. Man, there was a lot of dirt.

—JOHN YOUNG

After docking, open the hatch. God, the biggest thing I remember is the goddamn cloud of dust that they were operating in, in the lunar module. I opened that hatch and, Christ, I thought I was looking into a goddamn dust storm. . . . I made them take their damn suits off and hand them over so I could put them in bags right away. They were covered with that shit.

—DICK GORDON

When they came back, what did they do? The first thing they did was, they filled up my free space. . . . You can go down to the Smithsonian; there ain't much room in there. . . . Only time you could stretch out was when those clowns got out of the way, and then you could sit there and float around, and it was like a real spacecraft for a change. . . . And so, did you want somebody to come back and invade your space? And be dirty, too! Yeah, you get protective about that. You know, I got this thing all squared away, and it's clean now, and goddang it, I wanted it to stay that way. . . . So when people say, Weren't you glad to have those guys come back? Well, no, not exactly. I mean, I'm glad to see 'em, but did they have to hurry?

—KEN MATTINGLY

Because you were running on adrenalin, you didn't really feel the exhaustion. That all came after you got back in the command module, took off the suit and relaxed for a minute. And then you realized you were just dead tired. And we hurried. Even Houston hurried us, because they wanted to get us out of orbit on that next pass, because they knew we were all dog tired. As long as we had the energy going and the adrenalin going, to get the ignition of the motor out of the way, then we could relax.

—ED MITCHELL

When we got back in that command module, we were *home*. Never mind that we were in lunar orbit—it was home. You take off your suit and stretch out and wet a washcloth and wipe your face, just get the dust off. . . . And you don't think about anything until the next day, when you say, Oh, yeah, we've got some other things to do, and the most important of which is getting the hell out of here!

—GENE CERNAN

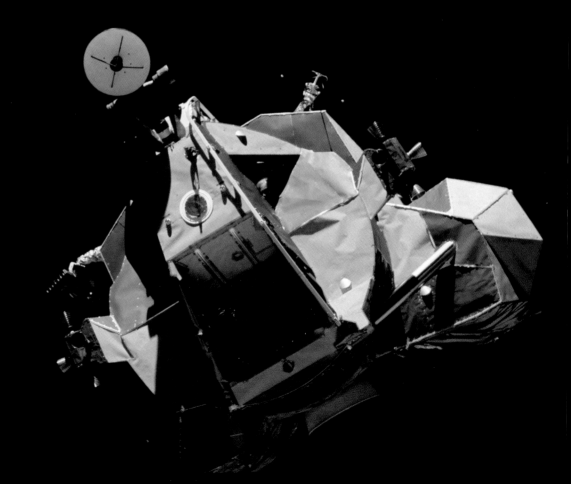

It's just like . . . looking back over your shoulder in an afterburner climb. You are *really* hauling the mail out of town.

—STU ROOSA

It's another one of those unbelievably spectacular sights. One of the few sensations of speed you can have in a space ship. You can— during the initial departure, you can actually *sense* that the moon is getting smaller. I mean, you can *see* it. It's like being in an airplane when you're really moving out, you can start to see motion. And leaving the moon, you can watch that sucker get small, right there in front of your eyes—I'm leaving that sucker! . . . And you leave with a sense of dismay. And you really don't want to. Because, I mean—it's really all over? I can't do this again tomorrow?

—KEN MATTINGLY

There's the moon that you walk on, and that you just touched down on, and that you approach for a landing. Then there's the moon from orbit. And then there's the moon from far away, whether you're in orbit around the Earth or not. . . . It really does appear that there are three different moons. One that you're on, one that you're in orbit of, and one that's a long ways away. And they don't blend, and you don't get the transition. There was only one exception to it, and that was shortly after we left the moon, heading back to the Earth, after we made the burn and it was okay. Then we could orient and look back and watch it grow smaller, from the back side. . . . And that was the only time when we were really allowed to experience the change.

—BUZZ ALDRIN

I accepted and realized the fact that, hey, we're not going to be back here for a while. . . . Most important is, You are here right now. You are ten thousand miles away from the moon, headed home. . . . And nobody's going to *be* here for a long time. Maybe for a long, long time. What should we be *doing?* Is there something we should say, or do, or be?

—GENE CERNAN

The lunar far side, seen from the departing Apollo 16.

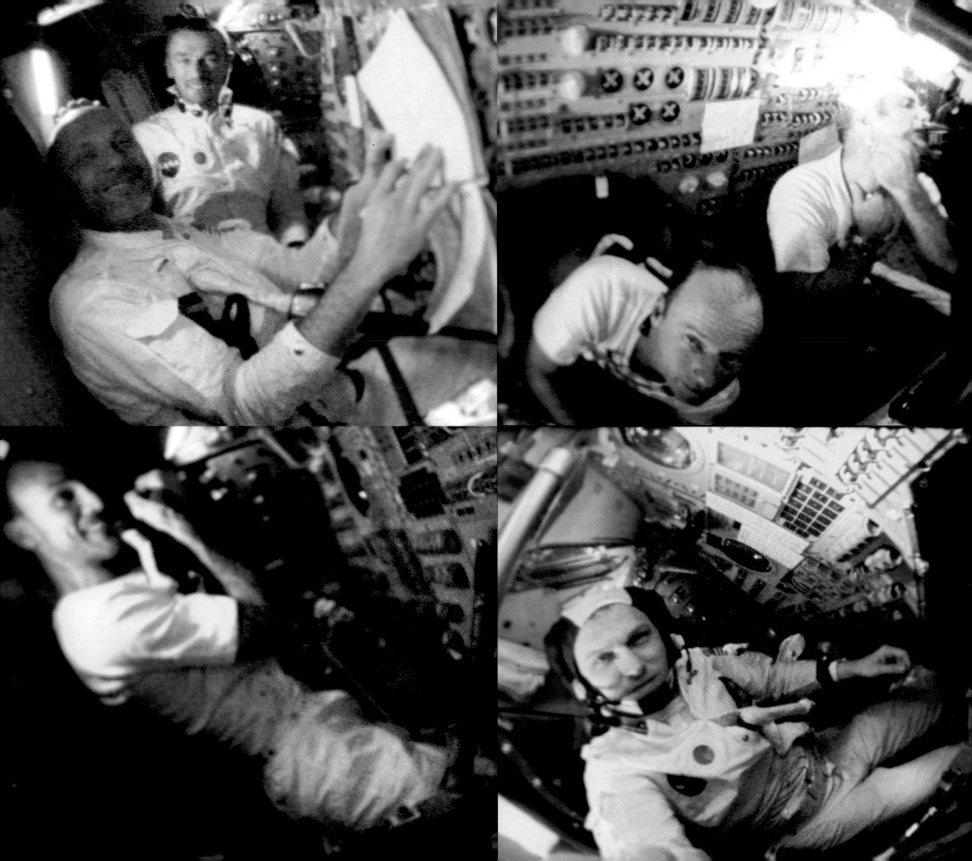

By the time you go fly, everybody wants things to break, so we can show you how *good* we are. As long as it works. . . . Don't give me more than can be recovered, but give me something really challenging, because I want to show you: Hey, babe, I can *handle* this. The guidance system goes out? No big deal! I can get home with a crayon and my pencil. I don't need any of that fancy computer. That's what you want to show. And you're disappointed when you don't get a chance to. 'Cause that's the level of proficiency you've trained to.

—KEN MATTINGLY

They surprised us with Christmas dinner. And I really got irritated when they put the goddamn brandy in there. You know, they had two little [bottles]. And I wouldn't let anybody drink that. . . . I just said, "Put it back. Because we're not going to take this brandy, and then have something screw up, and they'll say it's because we had the brandy." I thought it was a dumb idea. . . . People just don't understand—I was oriented to get the mission done. And . . . you could've gotten all screwed up. I think that's what the commanders are for.

—FRANK BORMAN

Relaxing on the homeward voyage, clockwise from above right: *Jim Irwin, Charlie Duke, Frank Borman, Ken Mattingly (exercising), Tom Stafford and Gene Cernan, Pete Conrad and Dick Gordon.*

You open this hatch, and you go outside, and it's dark. I mean, there's sun shining on the side of the spacecraft; you can tell that because it's—you can see it; it's illuminated. You know, and around the side it's black, because there's no sun over there, but I ain't over there anyhow, so it doesn't matter. You kinda walk down this rail, and you look around. And because you're going out in the sunlight, you've got this . . . gold reflective visor that keeps the stuff from coming in, and then there's this UV shield underneath that. You look around, and—I don't see anything. There's no stars out here. That's really strange. In fact, the only thing there is is this platform that I'm holding on to. That's all there is! *Nothing else! Nothin'.* I've never seen anything like it!

You had to turn your body to see [the Earth and the moon]. And you can, and there's a thing that's the size of [an orange], and that's one of them, and there's one over here, and it's a crescent, and it's not quite so big, but—that's all there is! . . . That's the only time I ever felt what it means to be in space. There's *nothing* around. Absolutely nothing. No up, no down. . . . The idea of up and down didn't mean a thing to me in the spacecraft. It didn't matter—I know how the instruments are oriented; I know where things are. You know, if I'm heads this way or heads up, it doesn't matter! I'm just kinda floating along; I do whatever I need to do.

And you get outside, and that was distinctly different. Because everywhere there was nothing. And I remember hanging on this rail, and the sensation hit me that there's *nothing* around here! And, man, I know I squeezed fingerprints in that rail. All of a sudden *(laughs),* I couldn't believe—This is really weird. Where did all the stars go? There ain't a star in this whole sky! I raised my [gold] visor, and that made me feel better. . . . *Whew*—There's stars out there! *(laughs)* You take the visor up and you could see 'em. It was just like looking out the window then.

—KEN MATTINGLY

I was having a ball. . . . You're not really a spaceman when you're in the confines of your spaceship. . . . You go outside, and you're hanging on, maneuvering out there from the safety and security of your mother ship. If you ever want to be a spaceman, that's the way to do it! . . . I wanted to do more. I wanted to go around to the other side of the spacecraft. . . . It was black on that side, I'll tell you. . . . I was trying to make a visual inspection of what was out there. Nobody else had done that. Here I'm up here, I'm an engineer; I'm doing my thing. It seemed to me like Houston oughta want me to go out and report, "Hey the color of this is . . . ," or, "I see a meteorite hole." Or anything. So what did I get to look at? The urine dump. "Yeah, there's yellow sublimated ice. . . ." Report that to Houston. Houston says, "Big deal, Ron, get back in there."

—RON EVANS

Coming back, you were still on a high. You had yet to get on the car-rier. You had yet to get home. A lot of things had to happen right. . . . The public adulation and the come back home and the whole thing—I mean, you were still headed uphill. . . . You're the world's hero at that point in time.

—GENE CERNAN

My thoughts of my family didn't begin coming back in my mind . . . until we started back to the Earth. I personally had kind of divorced the Earth. . . .

—JIM IRWIN

My biggest memory of the Earth is all white. I never saw that beau-tiful Middle East, unclouded area with continents and that sort of thing. Mine was always a crescent, mainly white. Because, see, by the time we came back it was just . . . a sliver of a crescent. . . . I think that might add to the effect, this feeling of how very, very small the Earth is. I don't know. Kinda hard to express. That's why we need to send a writer. We need to send a poet.

—STU ROOSA

A crescent Earth greets the returning Apollo 11 astronauts.

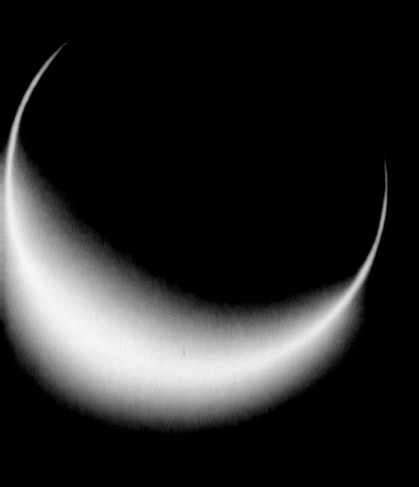

Now, we saw an eclipse coming back on Apollo 12, which nobody else had experienced before. It was a beautiful sight. It was very, very impressive. It looked like a damn diamond ring. . . . We were pretty close to the Earth for the eclipse, so it was pretty good size, but only a very small crescent of the atmosphere was illuminated. And then the sun kind of . . . went behind the Earth, and then came back out around the corner again. . . . The Earth was totally black; we couldn't see a damn thing on the Earth. Just black—I mean, nothing was illuminating it. There was the black of space, the black of the Earth that you couldn't distinguish anything on, and then we saw a little crescent of atmosphere. . . .

Before [the eclipse], there was almost a full moon, and we saw a very bright spot in the Indian Ocean. And we were trying to figure out what the hell that bright spot was. . . . It was a reflection of the moon onto the water that came back up to us, and it illuminated that whole area.

—DICK GORDON

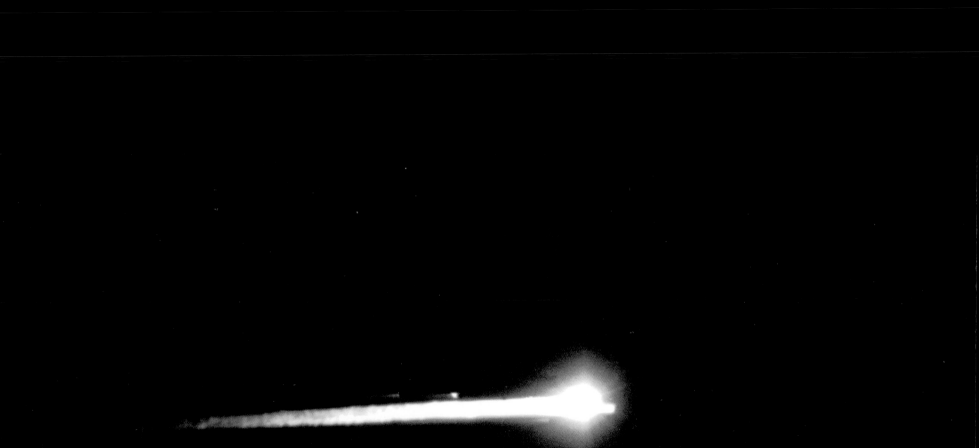

It's a very dynamic situation—a lot of monitoring, a lot of activity. And it's over very, very rapidly, too. You separate that service module, and you orient yourself to hit the atmosphere. . . . Christ, it's over in, what, eight minutes? . . . You've oriented yourself heads-down coming across that South Pacific, and you're really moving. I mean, you can really tell. . . . [During] translunar injection, you're oriented in such a way that you're really not seeing the Earth go by; you can't see it. But in reentry, you can. Coming back in, with your head down, you can look out that window, the side window, and you're really smokin'. Goddang, you're really moving. Those damn islands in the South Pacific are going by like—unreal.

—DICK GORDON

You sit in the simulator, and you practice reentries . . . and you watch the little G meter, you know, and . . . you imagine that you're going to pull a few Gs. . . . But that G force [on the actual reentry] is not an instantaneous spike, like it is on an airplane. . . . This thing, it lasts for 10 minutes! . . . You really notice the length of time that you're under these G forces, as you're decelerating. . . . That's the significant part, other than the fact that you see the fire out there.

—RON EVANS

Oh, that's spectacular. You are looking back to where you've been, and that heat shield burning off, it's a very spectacular sight. . . . You get the impression that you're in a corkscrew, because it's rolling. . . . So it looks like, when you look back, the ionized gas takes on a kind of a corkscrew pattern. . . . It's all kinds of colors. I remember some greens, pinks, yellows. Maybe red.

—DICK GORDON

The thing I remember about entry is being inside the flame, and wondering if it was going to scorch my back. I must say, I kept feeling I was getting hotter back there, but it really wasn't. But every now and then what looked like a large chunk would fly off, the size of your fist. And I thought, Jeez, you can't stand many of those. Later I realized that even the tiniest little grain, when ionized, makes a big ball of flame.

—BILL ANDERS

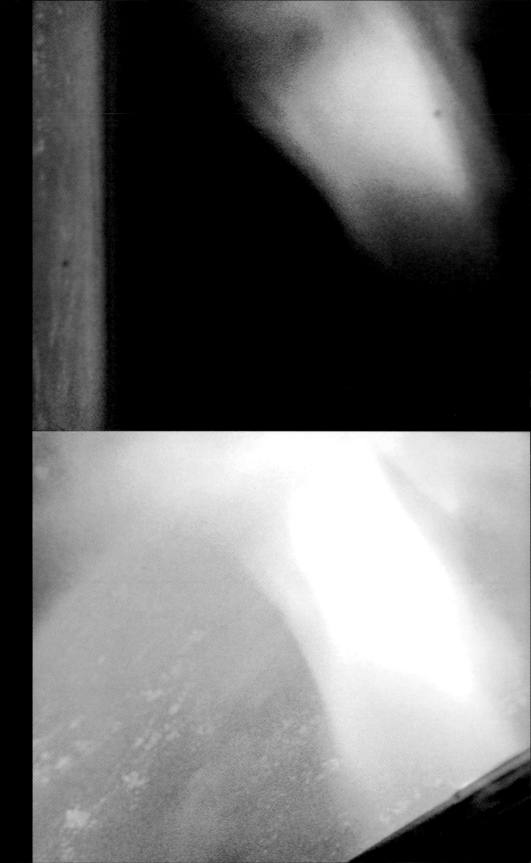

The contact with the water is a pretty good smash. As a matter of fact, the altimeter was wrong. . . . I was calling off the altitude to Gene and Jack. And we got down to five hundred feet. Four hundred feet. Three hundred feet—*Boom!*—we hit the water. So I wasn't prepared. And neither were they.

—RON EVANS

Only after I got back down in the Pacific and I was safe, did I have this thought: God I wish we were back there again. I suppose it's like *Romancing the Stone*. It's easy to tell about it afterwards and embellish it and say, "God that was fun. I wish we were back there." But at the time, it wasn't quite as much fun as you later realize it was. At the time, you're concerned about—hope we don't spring an air leak, I hope the fuel cells keep working, I hope Dave and Jim get off the surface okay, I hope we get home okay. We don't know that we're gonna get home okay until we get there. But once you get home, you look back on the quiet, the solitude, the peacefulness and say, "Yeah, wish I were back there again."

—AL WORDEN

Left: *Apollo 16 moments before splashdown.* Opposite: *Al Worden, Jim Irwin and Dave Scott enjoy the warm Pacific air while awaiting pickup by the recovery helicopter.*

When we got back on the carrier, and we played the tapes of the TV, reaction of the people, Cronkite saying, "Hey!"—whatever his reaction was. And just *(claps hands)* spontaneously, I saw that and was prompted to say something. I said, "Neil, we missed the whole thing." What I meant by that was, the three of us were more concerned with the event of landing men on the moon than any three men in the world. Because we were it! And yet, the response, the emotion, the satisfaction evidenced by these people watching that—we missed it! We didn't get to share in the exuberance of being in a crowd of other people and congratulating each other because "those idiots landed on the moon."

—BUZZ ALDRIN

There's probably a sense of pride, more than anything, and then you really got that when you got on the carrier: We did it. We did it well.

—STU ROOSA

Glad to be back. Period. Glad to see the family. And then, getting home, the neighbors had gotten together and built some kind of a sign that lit up, and the firecrackers were going off. And then being back in bed, snuggled up against Valerie, thinking, Did I really go to the moon? I mean, here we were back in the same old bed . . . and kind of wondering, Did that really happen?

—BILL ANDERS

At the end of every mission, every one was a letdown. You can only take adrenalin so far. Then you fall off the edge. It's almost like an intoxication, when you get that hyped up. Whether you're in the control center or on a support team, [or on the crew]. . . . The enormous intensity of devoting every waking moment to something—and then it's over.

—KEN MATTINGLY

Opposite: The Apollo 11 quarantine trailer is unloaded in Hawaii; (inset) the Apollo 12 crew greet their wives from inside their trailer. This page: (top) Gene Cernan and Ron Evans are welcomed home; (bottom) In quarantine, Ed Mitchell and Al Shepard inspect some of the rocks they brought back from the moon.

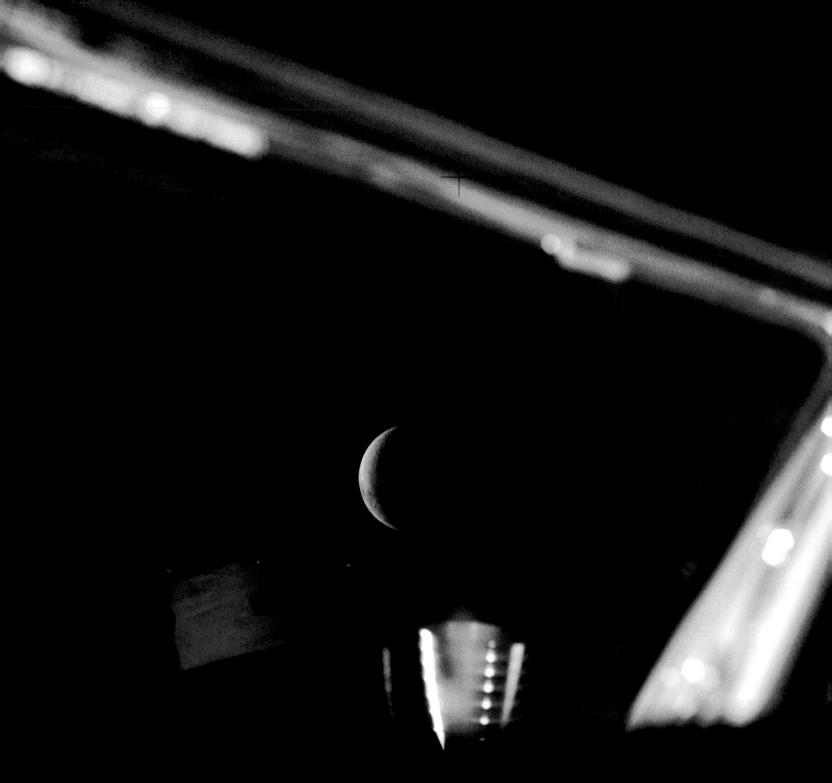

9

APOLLO 13

I knew instantly—not, frankly, how bad the situation really was—but for the
first indication that I could see, one oxygen tank was gone. . . . Which didn't
put us in a precarious state, but looking at the mission rules—and I knew it
already without looking—we couldn't even go into lunar orbit. So the mission
was gone, right there.

—FRED HAISE

The drama of Apollo 13 began even before the flight was under way. By early April 1970, mission commander Jim Lovell and his crew, command module pilot Ken Mattingly and lunar module pilot Fred Haise, had trained for almost a year to explore the moon's Fra Mauro highlands. When the crew was exposed to German measles, NASA doctors feared that Mattingly, the only one of the crew not immune to the illness, might become sick during the mission. Despite Lovell's objections, Mattingly was replaced by his backup, Jack Swigert, just days before launch. The trio left Earth atop a Saturn V booster on April 11, 1970.

Apollo 13 was a normal mission until the evening of April 13. At fifty-five hours and fifty-five minutes into the mission, when the docked command ship and lander were some two hundred thousand miles from Earth, the men heard a loud bang and felt the spacecraft shudder. In the command module, Odyssey, warning lights came on, and instruments showed that the craft's electrical system and oxygen supply were in serious trouble. It later became clear that an oxygen tank in the service module had exploded, crippling the command ship. The landing was immediately canceled, and the astronauts' mission became one of survival.

To make it back to Earth, Lovell's crew had to use the attached lunar module, Aquarius, as a lifeboat, harnessing its rocket engines, electrical power, and oxygen supply. It was too late to return directly to Earth; Apollo 13 would have to loop around the moon before they could head home. Before then, however, the men had to fire Aquarius's descent engine to correct their flight path. A second firing took place two hours after rounding the moon, to speed the homeward voyage.

With their command module completely deactivated, Lovell and his crew spent the journey crowded into the tiny cabin of the lunar module. With electrical power and cooling water at a premium, the men turned off nearly every system, and Aquarius's cabin chilled to 40 degrees Fahrenheit, while parts of Odyssey were near freezing. On Earth, mission controllers worked feverishly to plan rocket firings and other crucial measures, overcoming one obstacle after another to help the astronauts return home safely. Lovell, Swigert, and Haise splashed down in the Pacific on April 17.

You have to remember the makeup of the people that are doing this work; all were test pilots. Every time we get in an airplane, to test it, especially in any new regime, going out to any point, any new type he crew changeout was in itself emotional. . . . Just the trauma it caused, and again, you had a team, you had worked together for a long period, and now at the last minute, you're going to partially take the team apart. . . . Our normal practice was, several days before launch, we'd just literally back off. You know, you'd read your checklist a little bit, or go sit down at the beach house and relax, try to get a little rested, physically rested, before you got airborne. [But] we were in the simulators, if I recall, with Jack, to go through all the dynamic [mission] phases, to make sure we were talking the same language, until about eight o'clock the night before launch.

—FRED HAISE

It was inconceivable to me that they would leave me behind. . . . Turned the radio on and heard this goddamn news announcement that says, "NASA announces this afternoon that they've had a crew reassignment." That's when it became real.

—KEN MATTINGLY

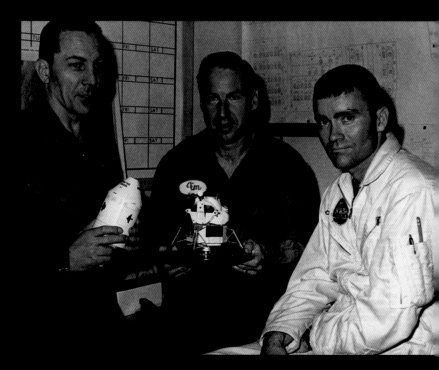

Opposite: Jim Lovell trains for Apollo 13's moonwalks. Top: Lovell with Mattingly (center) and Haise during recovery training. Left: Mattingly before a practice count-down. Above: Jack Swigert (left) replaced Mattingly just days before launch.

he difference between accidents in space and accidents in air-planes. . . . In an airplane accident, if the engine quits, or the wing alls off, things happen right away. In space, on Apollo 13, we had he explosion, and nothing happened. We were still going right along the way we were going. . . .

—JIM LOVELL

Most of the serious things I've had [in airplanes] were so quick that t was, frankly, only after the fact that, you've done whatever you've done, now you have time to even recollect and think about how bad it was. This, in comparison, was a slower-moving drama than any close midairs or several crashes I've had in airplanes. Nothing ike that at all.

—FRED HAISE

People often ask, "Did you panic when you had the accident on 13?" I say, "Well, we could have panicked, bounced off the walls for about five minutes. And when we finished up that, we'd be right back where we started from."

—JIM LOVELL

never felt we were in a hopeless [situation]. . . . No, we never had hat emotion at all. We never were with our backs to the wall, where here were no more ideas, or nothing else to try, or no possible solution. That never came.

—FRED HAISE

think that as long as we had an option, it [thoughts of confronting death] never really came up. . . . If there was a chance to get home, you work on the plus side; you don't work on the minus side.

—JIM LOVELL

We did not have a clear trail, from the time it happened, that I could see absolutely all of [the procedures] we had to evolve over the next few days, that the ground had to evolve. I knew they couldn't go to a cupboard and pull out a set of books that had all the answers. But I did not see anything present at any point in time that said they couldn't go maybe work it. In other words, if we had tried to activate the LM and it would not have activated, then we'd have been [out of luck]. Or I'd turned on the LM and there would have been nothin', no [cooling] water [for the electronics], all the water tanks had burst, right then I'd have known—we ain't gonna make it. But we had nothing like that ever occur. You understand what I'm saying? There was nothing there that said irrefutably we don't have a chance.

—FRED HAISE

And, of course, people often say, "Did you take a suicide pill?" or something like that. You didn't [need] those. All you had to do was crank open the little valve to the hatch there . . . open up the little vent valve. . . . Never would've thought about it until all hope was lost. And then our idea was, if all hope was lost, if we went by the Earth—say we missed the Earth, and we were on an orbit about the sun, if we had exceeded the escape velocity. . . . My idea was to hold off, you know, as long as we had options, as long as we could stand it, send back data. . . . We probably would have been farther out than anybody. And then, you know, then we would decide, you know, what to do. . . . Maybe we would have all committed suicide by opening up the vent valve. And that would have been the end of the deal.

—JIM LOVELL

The most impressive thing that I've ever seen was, Glynn Lunney walked [into mission control]. And if there was a hero, Glynn Lunney was, by himself, the hero. Because when he walked in the room, I guarantee you, nobody knew what the hell was going on. . . . And Glynn walked in, took over this mess. And he just brought calm to the situation. I've never seen such an extraordinary example of leadership in my entire career. Absolutely magnificent. No general or admiral in wartime could ever be more magnificent than Glynn was that night. He and he alone brought all of the scared people together. And, you've got to remember that the flight controllers in those days were—they were kids in their thirties. They were good, but very few of them had ever run into these kinds of choices in life. And they weren't used to that. And all of a sudden their confidence had been shaken. They were faced with things that they didn't understand. . . .

And Glynn walked in there, and he just kind of took charge. Restored everybody's confidence: Don't know what happened. Don't know where we are yet. That's our first job, is to figure out what our options are, and what we do. And we'll just get on with this thing. . . . At that point, nobody would even *think* of saying anything about disasters . . . it's just professionalism at its finest. That was all exclusively caused by one Glynn Lunney. Absolutely the most magnificent performance I've ever watched.

—KEN MATTINGLY

As soon as it happened—that was extraordinary. People just—within minutes, people started pouring in from all over. The next shift, people that weren't on shift. . . . Within an hour, the center looked like daylight. Everybody that had ever given a thought to Apollo was there. . . . We'll bring 'em home. I didn't know how. But I had then, and still have, absolute admiration for the group of people that ran those things. That says, by God, if these folks can't do it, then it requires violating the laws of physics, and they'll even bend those if necessary. But they can do anything. And there isn't any such thing as tired. There isn't any such thing as not enough. And . . . that was shown in Apollo 13. I mean, that was—that was the agency's finest hour.

—KEN MATTINGLY

I did not really think of the ground activity in that sense. I knew there were multiple teams. And again, I could not appreciate the exercise they had to go through in the development of some of these things. . . . Talking to a lot of the fellas after I got back, it caused them enough anxiety that even when they took off and went home, they couldn't sleep, because they were so excited. . . . And they were pretty fatigued, there was no question about it. But it was a side of the story we did not know in flight. In fact, the whole side of it on Earth was unknown to us. . . . We didn't even know how the Earth was receiving this.

In fact, we were very concerned that there was such a negativeness about this flight now being the first failure, and that bothered us, that we may be the cause of the end of the program. We really were bothered that that may be a possibility. In fact, there was Jim's one comment, which did sneak out live—about, it may be a long time before anybody comes [back] to the moon. That would have been, even more—in our minds, it would have bothered us a lot to be the cause of the end of a program. But everybody had been spoiled by such a string of successes. . . . You really cannot stop going forward because you have a failure. The only way you don't have failures in the flying game is, don't fly.

—FRED HAISE

Too many people always emphasize the ground's control of everything. Well, that's a bunch of BS. More than anything, [it was up to] the crew on 13—because we had no automatics, and we had no computer. We had no guidance system. You know, we didn't have a control system that was the standard control system. I mean, everything else was entirely different. . . . You flew by the seat of your pants. I mean, there was a case where you had to do your burns manually. You had to learn to fly the vehicle over again. . . . As far as spacecraft operations go, 13 was, I think, the epitome of man and machine, you know, trying to get back home safely. . . . A lot of the instructions on what to do came from the ground, obviously. Because they had all the people—the contractor people, the government people, the simulators—to check out those things. But to *execute* them—it's one thing to set up procedures . . . but to execute them when you know that your ass is on the line is a little bit different. And that is essentially what was happening on Apollo 13.

—JIM LOVELL

Scenes from mission control, clockwise from below: *Ken Mattingly watches the big screen; astronaut boss Deke Slayton* (right) *keeps vigil; an exhausted controller; controllers studying data.*

Was it a ground mission or a spacecraft mission? It was a ground mission; no question about it. God, you can't believe what those guys did. That was really, really impressive. Does that mean the crew didn't count? Oh, no. You have to have been in the situation. . . . Those guys lived in this absolutely, unbelievably *miserable* environment. And did what they had to do with a degree of concentration that they could avoid making mistakes. And it sounds like what they did was very simple. . . . But you can't imagine what it's like to do simple things when your body is so goddamn *miserable*. The goddamn cockpit was in the thirties. And wet. A hundred percent humidity. That kind of cold is penetrating. And if you can drink a hot cup of coffee, whatever, to warm you up, your spirits go up, you do lots of things. They didn't have any of those things. Everything they had was the same cold, miserable stuff. . . . And they were in an absolute survival mode with nothing exciting, nothing entertaining. Couldn't even build a campfire. And yet, they never lost sight of what was important. And they never lost their temper with each other, or with the ground. . . . If you had to have a commander to have that failure, Jim was the perfect commander: Mr. Cool—tolerant, patient.

—KEN MATTINGLY

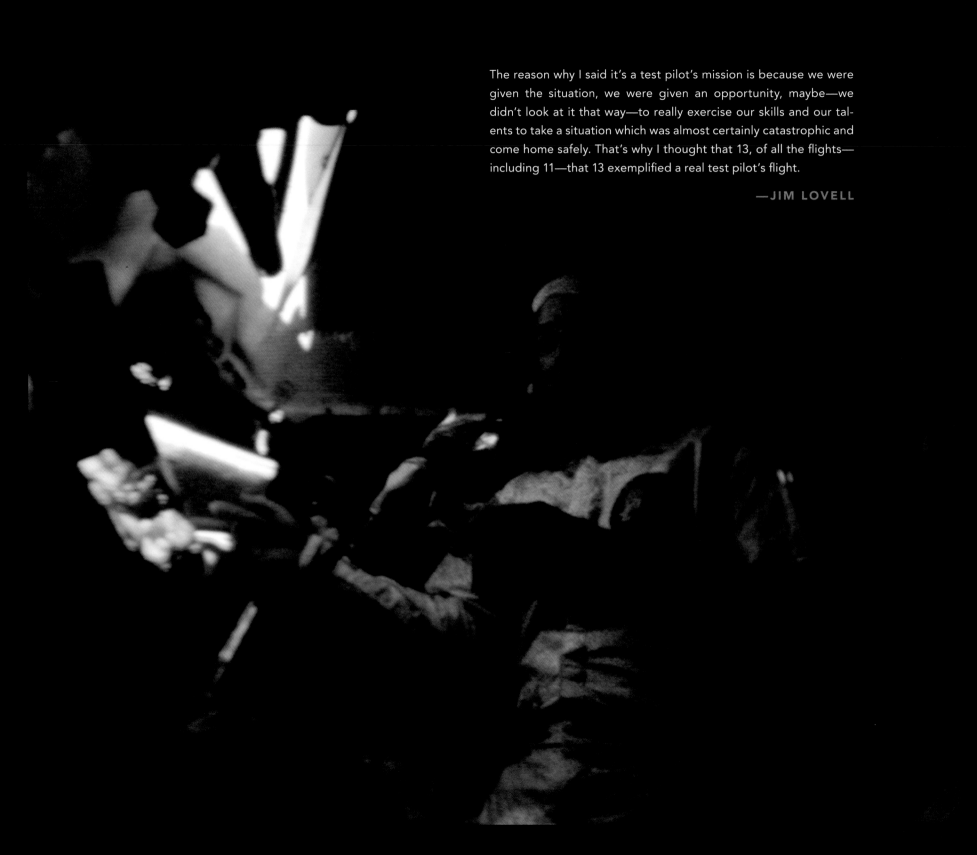

The reason why I said it's a test pilot's mission is because we were given the situation, we were given an opportunity, maybe—we didn't look at it that way—to really exercise our skills and our talents to take a situation which was almost certainly catastrophic and come home safely. That's why I thought that 13, of all the flights—including 11—that 13 exemplified a real test pilot's flight.

—JIM LOVELL

Our mission was a failure. I mean, there was no way around it. There's no question it was a remarkable recovery from a bad situation. But at the same time, relative to the mission intended, it was a failure. The biggest emotion I had for several months after that flight was disappointment. It was the biggest emotion in real time, when the explosion happened, was disappointment. Just a big sinking feeling. . . . Biggest disappointment of my life.

—FRED HAISE

Above: *Lovell and Swigert prepare for the mission's final, untried procedures.*
Opposite: *The damaged and now discarded service module drifts away from the*
docked command and lunar modules, shortly before reentry.

Our mission was a failure. I mean, there was no way around it. There's no question it was a remarkable recovery from a bad situation. But at the same time, relative to the mission intended, it was a failure. The biggest emotion I had for several months after that flight was disappointment. It was the biggest emotion in real time, when the explosion happened, was disappointment. Just a big sinking feeling. . . . Biggest disappointment of my life.

—FRED HAISE

Around the moon, when we were getting ready for this PC plus 2 burn [firing the lunar module's descent rocket two hours after the craft's closest approach to the moon] . . . these guys were interested in looking at the back side of the moon. I'd seen the back side of the moon. I'm trying to make sure that that thing is going to light off when we get started.

—JIM LOVELL

We really got out the cameras, at least Jack and I, and tried to make use of as much of the film as we could. We had several cameras loaded and ready. We shot a lot of pictures . . . just to document what we could. We figured, we may as well try and recover something. . . . We really didn't have anything to do in that pass, and we lost communication with the ground for a portion of that, behind the moon. And our next maneuver coming up was after we passed around the moon, plus two hours. And we were kind of ahead of the timeline in preparation for that maneuver. So we really didn't have much to do except to look, and shoot pictures. And it's a strange place. It's a big, beat-up rock. Not anything like on Earth, I'll tell you that. . . . There's nothing there. It's rock with a lot of holes in it.

—FRED HAISE

Rounding the moon, the astronauts had this view from the lunar module's overhead window.

Above: *Lovell and Swigert prepare for the mission's final, untried procedures.*
Opposite: *The damaged and now discarded service module drifts away from the docked command and lunar modules, shortly before reentry.*

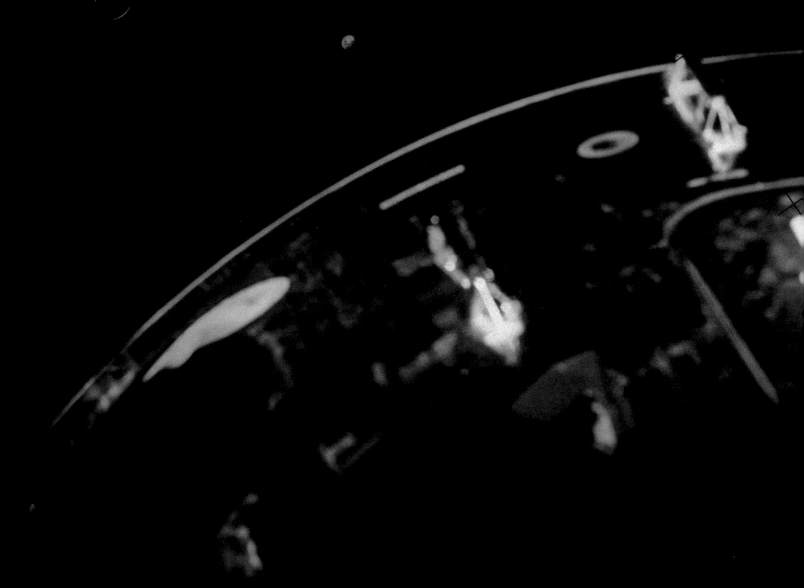

[Seeing the damaged service module] raised the, maybe, possibility of the heat shield being cracked when we saw that this whole panel had been ripped off the side. . . . It was a bigger thing, really, than we envisioned. We expected to see a little bitty ripped hole. I did, I should say; maybe Jack or Jim suspected more. I didn't; I was thinking I would maybe see one little gouged hunk of aluminum, and a hole.

—FRED HAISE

It is appropriate that there were *thousands* of people who personally felt that they made a significant contribution to getting those folks back. . . . There was this *unbelievable* amount of energy that went into making sure that from everybody's individual perspective, *their* part would work. And that's true. Every little piece. *Everything.* If anybody was associated with Apollo, their company was associated with Apollo in any way, the lights were on in their offices twenty-four hours a day for [four days].

And it was an extraordinary—one of the few times when this *huge* nation was unified with a single objective. The whole *nation* was. Not just us, but the goddamn globe. All over the world, people were concerned. Our people had the advantage of being able to be active and do things. The rest of the world could just share. But it was one of the most unifying experiences. . . . We finally have this situation where, no matter what your persuasions and political ideas are, for a time, there was this common goal that we hope these folks make it.

—KEN MATTINGLY

Celebrating Apollo 13's successful recovery: Above, Houston space center director Bob Gilruth; below, Deputy Director Chris Kraft lights a cigar for spacecraft program manager and former astronaut Jim McDivitt, with flight operations chief Sig Sjoberg; opposite, flight director Gene Kranz (right) takes in the celebration as Lovell, on the deck of the USS Iwo Jima, appears on the big screen.

We didn't realize, until we got on the ground, on board the ship—even the ship didn't realize the whole thing—of the amount of emotion and attention this flight had. . . . It was amazing, the amount of attention that it got. I was completely amazed.

—JIM LOVELL

Certainly one of the things that 13 did was to build confidence. Confidence in the contingency schemes and the interaction between the ground crew and the flight crew. You try to look at the positive aspects, and that was one of the very positive aspects of it. . . . The system works. You've had a catastrophic blowout of a major system, and you still bring the guys back.

—ALAN SHEPARD

The finest hour, in my viewpoint, of the space program was getting Apollo 13 back, not the first lunar landing.

—TOM STAFFORD

10

AFTERMATH

You come out of quarantine, and you ride on Air Force One or Air Force Two
. . . and you go to the White House, and you have steak dinners, and you're
on TV shows, and you go to Camp David. . . . Well, anybody that says that
life is the same before or after, just look at the facts. I mean, your whole life
changed. . . .

—STU ROOSA

remember thinking, I got a chance to go to the moon. And I can also remember thinking, It's going to change a lot of people, to do that. And I can remember wondering how it would change [those people]—and myself, the fact that I was going to do that. . . . I thought very hard about that I didn't want it to change me. . . . [And now] I don't think it changed me.

—PETE CONRAD

Space changes nobody.

—STU ROOSA

I was working for six years, dreaming to go up there and experience the zero G, and go to the moon, and have a great trip. . . . It was a beautiful trip, a beautiful trip. And I'm not going to say it's the same as riding on an airplane to Europe and looking at castles. But I have a good time on those trips too, see? Does that change me?

—RON EVANS

You get so much pressure, when you're out with people. They all want to meet you and share the experience. Sometimes you like to keep your own family removed from that so you can relax and be yourself. So there is a desire to be quiet. . . . I think there's also a responsibility to the people of the Earth to share the adventure that you've had. . . .

—JIM IRWIN

Return to Earth [Aldrin's book] was to readjust to life after an event that brought on tremendous impact, tremendous changes in a person's life. And it's not just that trip, because I've found there are a lot of little "return to Earth's." . . . Then you gotta come back and do the laundry, and all the rest of that stuff, and face reality. Sometimes as if it never happened. Or despite it ever happening.

—BUZZ ALDRIN

As far as returning to normal life—I don't think any of us ever returned to normal life; I don't think any of us were normal people to start with.

—ALAN SHEPARD

I don't know that you *can* prepare for that kind of thing. I think some people do it better than others, and I'm certainly not as adept as many in those regards. Nevertheless, that's a fact of life. That's the way that things fell, and you just do the best you can with them. . . .

—NEIL ARMSTRONG

You can never get rid of the fact that you were an astronaut. Sometimes it's a pain in the ass. The last thing in the world is to get on an airplane or be someplace, and here comes some guy or gal that's gonna beat your ear and *(emotional voice)*, "Oh, what was it like on the moon?" And usually if I tell 'em the truth, then they don't believe me. 'Cause they've got some preconceived notion that I should tell 'em I was frightened or I was awe-inspired or I saw the Lord or—I don't know. Whatever it is, it's not what they want to hear, that it was the right place at the time. That just shuts the door. Usually I tell them, "It was super. Really enjoyed it." That's the end of it *(laughs)*. And they don't want to hear that. I've had people get mad at me! They get mad at me because they think I'm insulting them.

—PETE CONRAD

We flew [Apollo 17] in December of 1972. The POWs were released from Vietnam in January of '73. We happened to be at Frank Sinatra's place in Palm Springs. . . . Here we are, watching friends of mine as they walk down the ladder after they landed back at Clark Air Force Base. They'd been POWs, some of them, for six years, and some of them from before the time I was over there. They didn't even know that I'd gone to the moon. . . . But for the grace of God, I could have just as well have been in the Hanoi Hilton. And instead, I'd been selected into the program, gone to the moon, and returned the month before they did. I'll tell you, that gets to you.

—RON EVANS

Mike Collins, Buzz Aldrin, and Neil Armstrong engulfed by admirers in Mexico City during their postflight world tour.

I've been really happy since the moon trip. I've felt that I got a lot of luck in life to be able to do that and go to the moon. And that I shouldn't really complain about other things. I mean, I've had all this tremendous luck. . . . That really, I ought to be pretty thankful about it. You know, if it rains tomorrow, so what? Look at this other wonderful thing that was given me. So, it's made me real content, real happy since then. I'm really a contented person, and . . . I feel good every day.

I remember thinking in lunar orbit, that if I got back from this, I was going to live my life differently, in that I was going to try to live it . . . *like I want to live it*. I'm just going to do like I think is right. That's what I'm going to do—I'm going to do that more, because I don't want to miss this chance. . . . Mostly it made me have a lot of courage to do what I wanted to do and be happy about it. . . . That's one thing that really allowed me to be an artist. I probably wouldn't have had the courage to be an artist.

—ALAN BEAN

It's my belief, no one could've participated in such an adventure without reinforcing your belief, in whatever belief you had before you left. . . . In my case, it confirmed there is a God.

—RON EVANS

We had a chance to reflect on the mission when we were making that flight [after splashdown] from Hickam Air Force Base in Hawaii back to Houston. It was a long flight. . . . We knew we were contracted to the New York Times for our personal stories. So it was a good time for reflection on where we'd just come from, what we'd done, and what it meant to us. And that's when I realized that the mission had great spiritual meaning to me, spiritual impact. So I added that to my report. . . . Any time we traveled after that, I talked about the science and the technology, but also, if it was appropriate, add something about the spiritual side. . . . When I came back, my new purpose in life was to share faith, a faith renewed, a faith restored. A faith that came alive on another world.

—JIM IRWIN

There was a vague feeling that something was different. That my life had gotten very disturbing, very distressing at a subconscious level. And the postflight period . . . and the hero plaudits, and the parades, and all of that stuff should have been The Return of the Conquering Heroes. It should've been great. It wasn't. Oh, you know, it was very nice. But I wasn't there. There was something nagging, troubling. . . . And it wasn't until years later and as I got further involved into these things that happened after I left NASA, before all of that started becoming clear as to what all this was about. . . .

What I do remember is the awesome experience [on the trip back from the moon] of recognizing the universe was not simply random happenstance. . . . That there was something *more* operating than just chance. . . . I've assiduously spent the last fifteen years figuring out what was true. . . . I know that information flows . . . [and] transmits in some way beyond electromagnetic equations. . . . I see [the universe] as a learning organism, like the human is a learning, growing, changing, information-assimilating and -organizing organism.

—ED MITCHELL

I think, from within, I think it gives you some confidence. I think if you can handle it right, you can decide that, hey, I have to stop trying to climb for the top of the mountain, because I've already been there.

—STU ROOSA

We all had something we wanted to keep hidden from someone else, certainly any weakness or lack of knowledge. You always wanted to be a super achiever, super knowledgeable, super in everything. . . . I knew that really wasn't true—who was I trying to fool? I was fooling myself. . . . Maybe that's why I welcomed the opportunity to reveal my humanity, my normalness. That I am like others. That we're not a superhuman breed. . . . I welcomed the opportunity to drop that veneer, that cloak of secrecy, and really show that, man, I've made mistakes. . . . It became maybe a relief for me to reveal myself as not the perfect human being that they thought astronauts were.

—JIM IRWIN

I can remember, shortly after we were back, we had a session at the United Nations. And we addressed the General Assembly. And I made the comment that, if everybody in this room had just had the view of Earth that I had, that the discussions would go a lot smoother. And I think that's true.

—STU ROOSA

When I came back from Apollo 8, my mother, who lived about fifty miles up from Cape Kennedy—I told her all about the flight and everything like that. Now, she was seventy-something at the time. She could really hardly believe that I went to the moon. . . . You know, she listened and everything, and all that. But I could tell that, you know, really, I mean, *going to the moon*. And my son, my young son, the one who was born after Gemini 7, let's see, he was three years old, after Apollo 13. . . . Anyway, going to the moon was no big deal for him. Because as long as he could remember, people were doing it. And that was history to him, as he grew up. Whereas, my mother, you know, she was born in 1895, and, you know, the moon—going to the moon, you might as well have gone to Saturn, or something like that. She really couldn't comprehend that we actually went to the moon. . . . What impressed me was, in just two generations still living, the tremendous comprehension of what we had done—hardly believed in the older generation; no big deal in the younger generation.

—JIM LOVELL

This page: *Jim Lovell* (above) *with the Apollo 8 command module, 2005. His crewmate, Bill Anders* (left) *in his 1956 Broussard airplane, 2002. Opposite: Earth from Apollo 8.*

The biggest philosophy, foundation-shaking impression was seeing the smallness of the Earth. . . . Even the pictures don't do it justice, because they always have this frame around them. But when you . . . put your eyeball to the window of the spacecraft, you can see essentially half of the universe. . . . That's a lot more black and a lot more universe than ever comes through a *framed picture*. . . . It's not how small the Earth was, it's just how big everything else was. . . . You look around and you say, "*Shiiiit,* there ain't *anything* else, anywhere." I mean, that is *all* there is, really, except for the sun, which you didn't look at. You look around, and you don't see any stars; you just see this dull black, and this [Earth] is the only thing there. Okay?

And that's why I don't think we've ever really gotten it across to people through the photography about what I call the *perspective* of it. That you've got to see all the black, all the nothing . . . in order to get totally appreciative of the smallness, *aloneness,* insignificance of this pretty little ball you're looking at. . . . And so therefore in my talks, I've tried to make that point, that it really is a very big place out there, and we really are a little dust mote in space. And use the analogy that, [if you go] two times [farther out] from the moon, ten times from the moon, you're hardly anywhere. And pretty soon, you can't even see the Earth. . . . That this planet that we're on is *puny*—in a physical sense.

But in contrast to that, in complicated contrast . . . it is important in a human sense. . . . You get this contradiction. . . . Physically insignificant, personally important. . . . And try to get people to think that, sure, Earth is important, it's our little originating habitat. On the other hand, on a galactic point of view, it's nothing. So we get all hung up on whether, you know, somebody's going to have a leaf fire down the street and a zoning ordinance or something like that; those are really kind of small-potato considerations. . . . It's too bad we can't pull ourselves up out of looking at life right at the end of our nose, and try to take a broader view.

—BILL ANDERS

I cannot count the number of times somebody has said to me, "What does it feel like to be on the moon?" I'm tired of that. . . . I don't want to try and figure that out anymore. I've done my best.

—DAVE SCOTT

When I came back, I got so many questions about, "How did it feel? Did it change your life? How do you view Earth?" You almost have to manufacture a response to that question.

—CHARLIE DUKE

I enjoy being a hero, because that's an easy opener for me to portray the humanistic aspects, to answer everybody's question, "What is it like in space?" It really turns me on. A lot of people in this country helped me get up there and do what I did. And there's a tremendous excitement that comes back to me when I get to speak to people about what it's like.

—RON EVANS

I'm sure every explorer had the same feelings. . . . Lewis and Clark, how do you describe the Columbia River or the Rogue River? You didn't even have photographs in those days. How did the Spanish conquistadors describe the Aztec Indians? "You wouldn't believe this place—it was all gold!" You can't comprehend—people, the human being, has to, to me—to live it before you can comprehend the beauty, or the effect of something.

—CHARLIE DUKE

I'm not sure really how the layperson reader ever is going to grasp whatever words are going to try and describe this. I've felt totally inadequate in ever trying to do it with spoken words.

—BUZZ ALDRIN

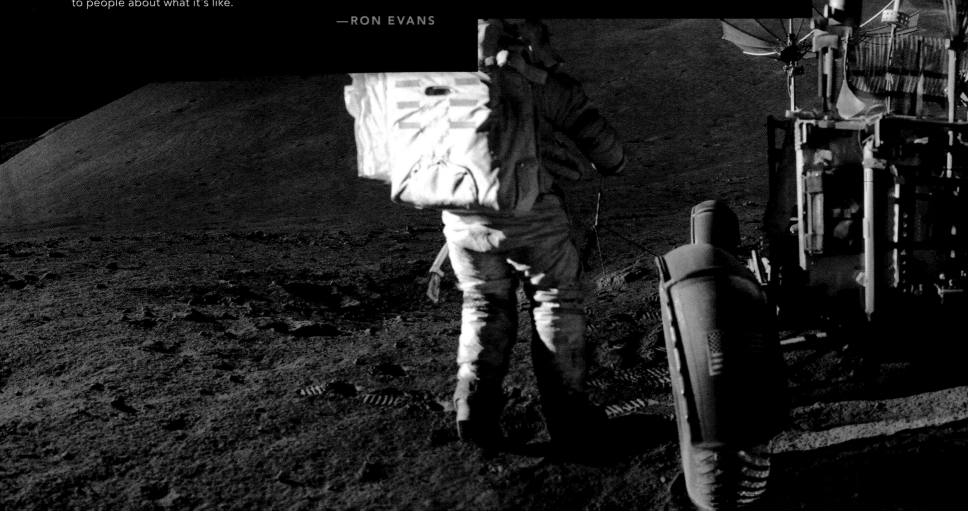

We didn't know how to talk about our feelings. I didn't know what feelings were. When I used to give lectures, people used to ask, "What did it feel like to be on the moon?" I didn't know what it felt like! I could tell them what I did, and what I thought, but not how I felt. It pissed me off.

—ED MITCHELL

I feel I'm different than the other astronauts, and I know that sounds like an arrogant thing to say; I don't mean to be arrogant at all. But by that I mean I have written—by myself, not "as told to"—but I have sat down and sweated through two books [*Carrying the Fire* and *Liftoff*]. And I have, for better or for worse, convinced myself that I have said everything I want to say. . . . No, it [public demands] hasn't ruined the experience for me, but on the other hand, having written the two books, I feel like I've said everything I want to say about the space program. . . .

The unfiltered sunlight of space illuminates Dave Scott and the rover during Apollo 15's first moonwalk, with Hadley Delta mountain and Hadley Rille in the background.

I have had it with certain questions, and maybe the way they're asked. But I've learned; I've gotten slyer. I've gotten more adept at just deflecting them. I mean, if someone says, you know, "What was it like up there?" You know, I've just written this fucking book telling 'em four hundred pages of what it was really like up there, and they say, "Oh, I loved your book! Now tell me: What was it really like up there?" Well, I learned to deflect that. It depends on their age. I mean, if it's a young kid, you just say, "Oh, it was cool, man." They say, "Was it?!" "Oh, shit, yeah. Oh, it was cool." And they walk off; that's great. It's [that] they want to hear the sound of their own voice, or they want to somehow interject their own question or their own inquiry into the process. And that's all right. So I go by the age; I usually give 'em a one- or two-word answer, depending on how old they are. And they usually—sometimes, though, they won't be satisfied with that, and they'll want to go on. But nine times out of ten, they'll say, "Oh, oh, uh-huh." And they're very pleased.

—MIKE COLLINS

All we can use is analogies. What do you want? There are only analogies. It's like the moon. It's dramatic like the moon. You say, "How is it dramatic?" It's like the moon.

—DAVE SCOTT

think you all overemphasize this emotional, what-does-it-do-to-the-psyche? and all that sort of stuff. We're engineers, we're test pilots. We're not doing anything different from taking up an airplane. And going out to the moon is certainly a first, and it's awe-inspiring, and it's great. But I mean, there's no great emotional change. Nothing in zero gravity or space changes anything you think, or anything like that. . . . I mean, it's just another extension of our exploration.

—JIM LOVELL

I've always said that the thing I did the best and enjoyed the most was getting there and getting back. The fact that I had to do some work when I got there—it wasn't "exploring," it was payment for getting to make the trip *(laughs)*.

—PETE CONRAD

Only thing was that the moon was not that far away anymore. You know? Someplace I had been to. Takes less than three days to get there. . . .

This is a big misconception by people who haven't gone to the moon. I mean, it's like going to Indianapolis—or a little bit farther, but I mean, I don't envision going to Indianapolis, you know, as something that's admired. Going to the moon is different, but I had the proper equipment to do it. I had the proper transportation system. And it took three days; going to Indianapolis only takes a couple of hours. But, I mean, still, the tools were there. The ability was there. We did it. . . . There's *nothing* mysterious about it. There's nothing mysterious about spaceflight.

—JIM LOVELL

People don't understand—they really don't understand, the average Joe on the street . . . that this isn't some great fucking experience that, you know, is mystical, magical, changes your whole life, Jesus Christ, this, that, and the other thing. It's just a fucking pile of rocks that happen to be 250,000 miles away! It was a real challenge and a hell of a lot of fun to get there. But after it's all over, you can't say that that's where you want to retire because it's pretty. It's not. It's beautiful in its starkness, but you don't want to sit there forever and look at gray rocks! Or brown rocks, whatever.

—PETE CONRAD

The cost of getting to fly to the moon: Pete Conrad exploring the Ocean of Storms.

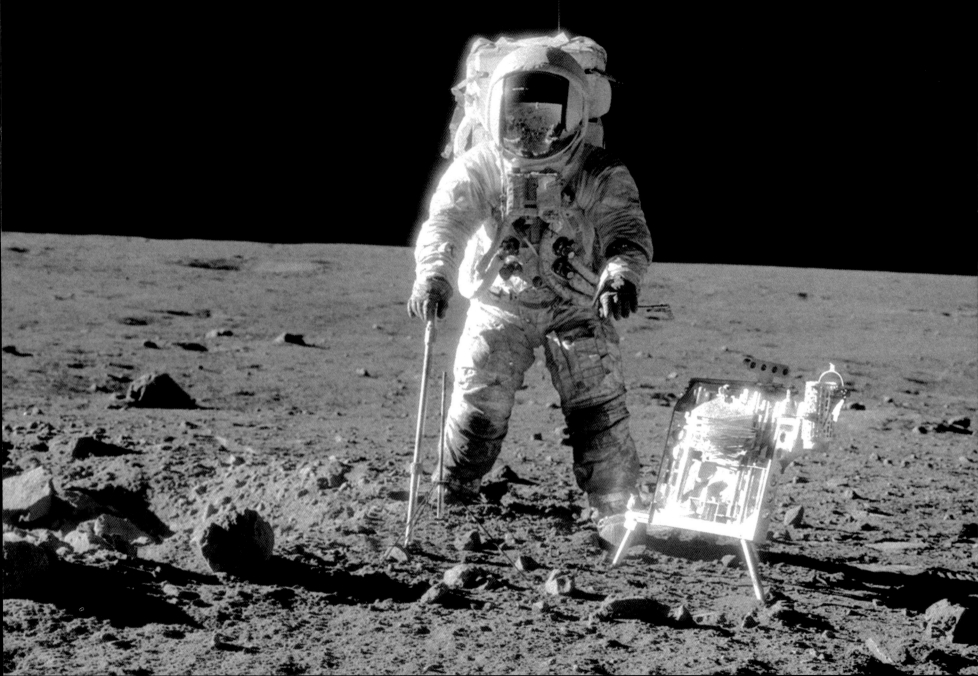

Going to the moon is an extraordinary thing! It doesn't matter if you're *first* or *last*; it's extraordinary. Now, to the public, first means something singular. You know, but to go to the moon, I don't give a shit if I'm the *last* guy, as long as I get to go. I'd *rather* go first. But I'd rather go last than not go. Right?

—KEN MATTINGLY

If I had hindsight, you know, say, okay, let's turn back time. You pick the flights you'd like to be on. I would've liked to have been on two. . . . I would've loved to have landed on the moon. But if I had another choice in going, I would have picked 13 all over again. . . . You know, I'm disappointed that we didn't land on the moon. But to put yourself in a position that you're able to get out of . . . to show the depth of expertise, both on the ground and in the air, and to bring back an almost catastrophic situation, is the real test, I think, of a test pilot.

—JIM LOVELL

I'm often asked that question because it's a very obvious question. When you're out here visiting . . . the Lions Club or something, they say, "You got so close, but you didn't get on the moon." But they have no concept of *one,* how proud you are to be selected, and *two,* that you're on a lunar crew, and *three,* that it's a landing and so forth. . . . I say no, I don't [mind], but they don't believe you. I don't know of a good answer to that. . . . You can say, Twelve people have stood on the surface of the moon and looked at the Earth. Well, and there have only been six people that have orbited the moon solo. There have only been six of us that have viewed the Earth from the lunar distance by yourself . . . so we've got a smaller fraternity, if you want to count numbers.

Oh, yeah, I cared. I never knew anybody who aspired to be a brides-maid. I mean, maybe somebody does, but it sure as hell wasn't me. I mean, you go to the moon, you go walk on the goddamn moon. You don't go around and look at pictures of it. . . . If you fly on a landing mission, you gotta feel shortchanged that you didn't get to land. But then you say, "My God, I've had a chance to do things that nobody ever gets to do! Who am I to feel sorry for myself because I didn't get to have this even *more* extraordinary dream?" You can't bring yourself to bitch. But that doesn't keep you from feeling sorry that you didn't go all the way. . . . Yeah, I missed that. That's what I wanted to do. I'm not *complaining,* but you know, did I set out to go walk on the moon? Yeah! Did I get there? No. Am I disappointed? Yes! Would I give up being second for anything? No, of course not!

—KEN MATTINGLY

I don't focus on self-satisfaction, and being first, and those kinds of things. . . . I take certain pleasures in the achievements, the techni-cal achievements. And not just in an overall sense, but little details here and there. Finding ways to accomplish the job. Those are the satisfactions I got. . . . Those kinds of little things that individually don't mean anything, but collectively—not just those, but hundreds or thousands of little things that you're involved in over a period of years that built all the little components that allow you to put together a car. . . . And you take great joy out of the time when you come up with some new gimmick or procedure or gadget or approach that allows you to get a little further than you ever were before. Get to a new plateau to look from. That's just the way I tend to look at things.

—NEIL ARMSTRONG

When I landed on the moon and came back from the moon, I was thirty-nine years old. My career had been finished. I'd finished my career. That's it. Now go find a new career.

—DAVE SCOTT

It was a great experience that I had. It's one of those things that comes along, not once in a lifetime, but once in a millennium, maybe.

—DICK GORDON

You know, what does the pro football player do after he bungs up his knee . . . and he's twenty-eight years old? It's that sort of thing, and that's certainly a negative, having the most interesting job of your life, in a sense, peaking out at age thirty-eight. I think if you wanted to plan a moon flight, you'd probably want to plan it for when you're fifty-five, or something like that (laughs). And that's a negative, I think, to go from thirty-nine, forty, on—and this knowing you're never in your life going to have a job as interesting as that one again. But again, I would try to put it in a perspective of saying [it's] a small negative versus a very large positive. I mean, I don't resent having peaked out at age thirty-nine. I don't think I've—I may have done not quite as good a job on some of my subsequent jobs, because I—possibly I have in the back of my mind, Well, the stuff I'm doing now just isn't as important as the stuff I was once doing. But it's a small negative compared to the positive of having that recollection. And it's nice to go through life with people saying, "Jeez, that guy flew to the moon."

—MIKE COLLINS

I'd go back [to the moon] tomorrow—I thought it was a super place—and continue to work there, if we were going to do that. I would've stayed to do that. But that wasn't going to happen. So when that left, why should I look back?

—PETE CONRAD

We all had the best job in the world. We really did.

—RON EVANS

I like to reflect back and reminisce about the fun, and the adventure and the friendships. That's all a great part of my life. But what I can't do is look back and say, "Fifteen years ago I walked on the moon and I had done it all." You can't live a full life on an adventure that happened fifteen years ago. I can't look back and say, "I'm complete! I'm whole! I'm full! I did it. Fifteen years ago." It just doesn't work that way.

—CHARLIE DUKE

It still amazes me that I was able to do it. . . . You have no comprehension, from a standpoint of—if we went from this town [I grew up in], and drove eighteen miles to the Kansas border, that was a weekend drive. . . . And then it's totally incredible when this same person, where it was a weekend trip to drive eighteen miles, leaves the Earth. . . . It's incredible, that whole journey from Kansas to the moon.

—RON EVANS

Not even a moonwalk is life to its fullest without God.

—CHARLIE DUKE

You know I'm not kidding when I say there's just going to be nothing to replace that in my life, ever again. I mean, there's nothing I can [do] that would give me that much satisfaction. A lot of things will come close, but. . . . And you can't dwell on that, because if you get fixed on that, then you're not any good for anything else. You see what I'm saying? I mean, life has to go on . . . you've got to define new challenges.

—ALAN SHEPARD

It is difficult. Nobody's going to get over it. There are those of us who think we have, and nobody has.

—STU ROOSA

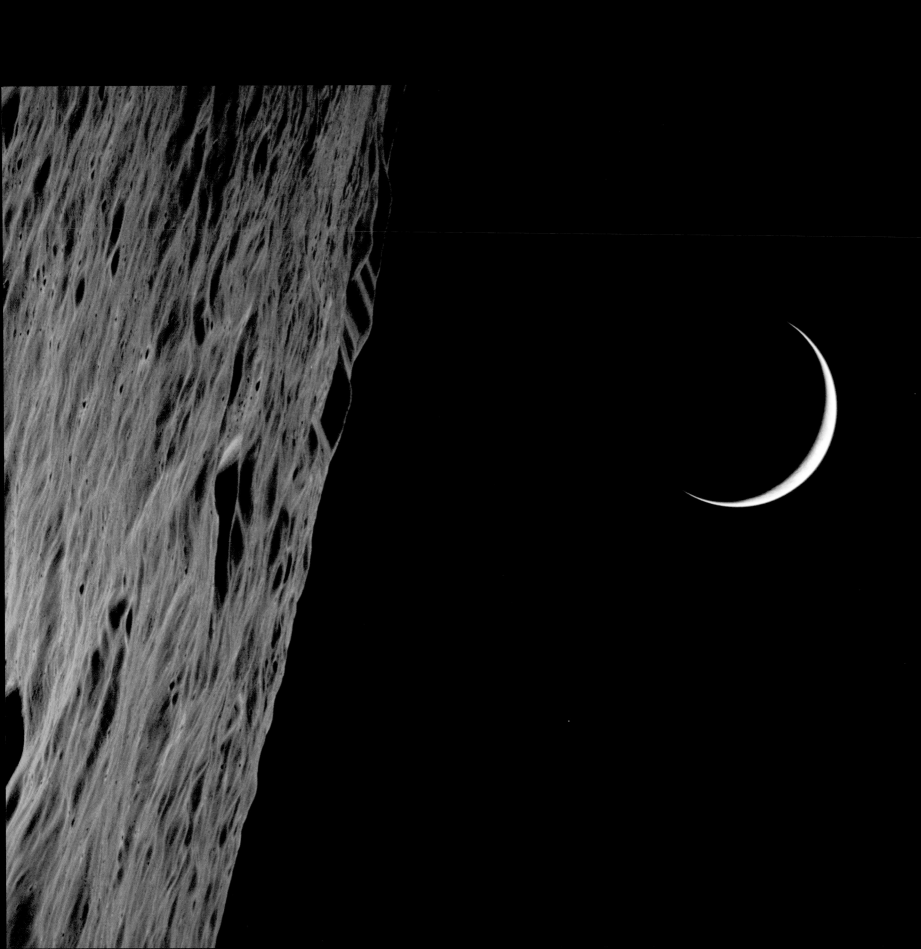

11

REMEMBERING

I have done things and been places you simply would not believe, and I keep that inside me.

—MIKE COLLINS

I hadn't been in the air an hour, and I knew I was in deep trouble. Because my mind was being overwhelmed with one extraordinarily impressive view, image, picture—and as soon as you got that and you said, *"Look at that!"* And then, *shit,* here came another one that was even more impressive! And I remember we were hardly out of Earth orbit when I said, you know, I'm in trouble. I can't—I'm gonna—I've got days of this ahead of me, I'm gonna forget all this stuff. There's only so much memory. And if they keep stuffing this memory into my brain, shit, stuff's going to run out the bottom, and I will miss it! And I'll never get it back.

   I remember that vividly, because it was this feeling of despair that, I can't *write* fast enough, I can't *talk* fast enough. I can't take *pictures* that are going to capture this. But, shit, somebody ought to tell the world how absolutely spectacular this stuff is! *(laughs)* And I'm gonna forget it *all,* 'cause tomorrow there's going to be something even *more* impressive! And it was; it was that way for ten days. Every time you turn around, shit, there's this new image that overlays the old one. . . . It was really frustrating. . . . For ten days it was this kind of stuff, with one image overlaying the next, and you just knew that—I don't want to look. It's going to screw up my memories *(laughs).*

—KEN MATTINGLY

Ken Mattingly snapped this picture of Earth shortly after Apollo 16 headed for the moon.

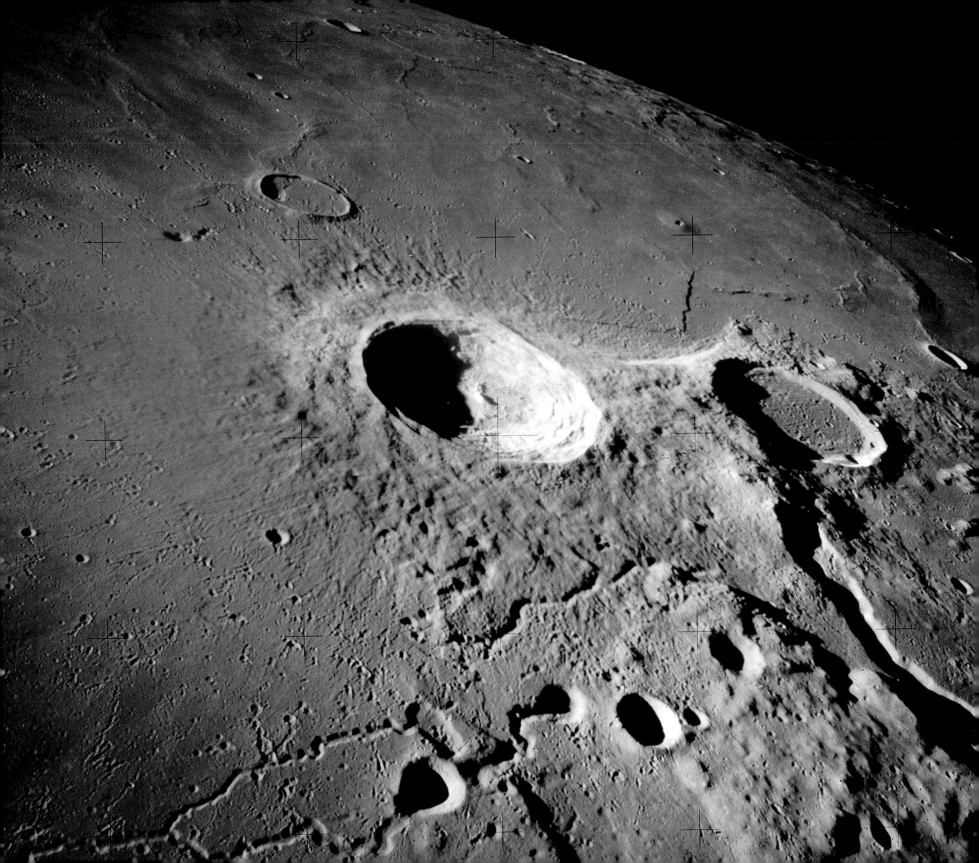

I was thinking this morning—you really can't afford it—I mean, it takes two hours to make an orbit of the moon—but I sure wish we'd have the luxury that you could just point the spacecraft down and coast one entire orbit of the moon without any experiments, without any radio transmissions, without any updates, and just spend one entire orbit doing nothing but just looking at the moon.

—STU ROOSA

The brilliant crater Aristarchus (center), flanked by the winding Schröter's Valley *(right), photographed from Apollo 15.*

People don't remember that Apollo 14 had a lot of great things in it. It was a *hell* of a good flight. Someone says, "Oh, yeah, you were the third to land on the moon—but didn't you play golf up there?" I probably should've known that was going to happen.

—ALAN SHEPARD

I was going to, on 13, on the way home, on one of the TV programs, was going to put some lunar dust in my mouth . . . and eat it, saying that this stuff is, you know, very pristine, and there is nothing wrong with it. . . . And hopefully in twenty-one days, I didn't break out in some rare disease of some sort. But I was pretty sure that nothing was going to happen.

—JIM LOVELL

[Dave] went off about three hundred feet to the east of where the lunar module was, to park the automobile. That's when he put out the memorial plaque and left the fallen astronaut [figure] there, as a salute to all the guys, including the Russians, who had given their lives for the exploration of space. And so that took a little while. And in all of our good planning *(laughs),* we'd never figured out what I should be doing during that time. I could be doing a little more geology, I suppose; it would be my only time to really do any real geology on the moon. But I just figured it was my time to relax, so I just ran around the lunar module a few times, and tried to do leaps across craters. And just felt like a little kid at recess time.

—JIM IRWIN

I would have loved to, on Apollo 13, had I landed, to land and discover the remains of an old Martian spaceship. And the skeletons of some old Martians that didn't make it. Wouldn't that be great? . . . I mean, if you brought back something that was entirely different, you know, some evidence of some other people *being* there before, or the remains of an old camp of some sort. . . . That would have been *really* exciting.

—JIM LOVELL

After the second page [of my cuff checklist], there's this naked broad. . . . Not a hell of a lot I could say about it over the radio.

—PETE CONRAD

When I'm thinking to myself about the most interesting memories or the ones that I like the best, they include Pete and Dick. But then, I never tell those stories, because they're sort of—I've probably told them to you—but I never otherwise bring 'em up, because they seem a little mushy. . . .

—ALAN BEAN

I collect hats. . . . The other thing, which didn't make the flight, which I have at home, is a gigantic goddamn hat, because everybody knew I collected hats, that I was going to wear on the lunar surface. But it didn't make the flight. It's a gigantic one to fit on top of the helmet. It was a blue and white ball cap, and it was all made to fit on top of the helmet. Somehow, the guys—It was fairly big, so I don't think they figured out how to get it on the spacecraft and get it by the inspectors.

—PETE CONRAD

Above right: *Conrad, Gordon, and Bean during training.* Right: *Pete Conrad, sans baseball cap but with R-rated cuff checklist, on the Ocean of Storms.* Opposite: *Relaxing on the homeward voyage, Alan Shepard makes a show of offering a razor to a bearded Ed Mitchell, who refuses it.*

Without telling anybody, we were going to take a camera off, screw it on the pole [on the tool carrier], put the timer on it, stick it in front of the Surveyor, and both of us were going to go over and have our picture taken. And we knew that's the one they were going to release. Until somebody says, "Who took the picture?" . . .

I got the [self-timer] device. Smuggled it on board. It wasn't very big. . . . We went out on our second EVA and put it in the bottom of the rock sample bag. And we went about that traverse, all around there, and got down to the Surveyor. . . . And they told us that it was time to take a little rest. And that was our cue to . . . get the picture. But we obviously couldn't talk to one another over the air, or the ground would know what was going on. So we were doing this by hand signals. So Al was carrying the rock sample thing, and it was full of rocks . . . and the blivet was down at the bottom. . . . So, Christ, he's holding that, and I'm pulling rocks out. We didn't have suits that you could bend over in, so I didn't want to put them on the ground. You know, after a while, we were working pretty hard. And Al's holding the tool carrier up, and I'm holding rocks, and he's holding rocks and the tool carrier. And the ground keeps calling us, and they are asking us if we're resting. And we're saying, "Oh yeah, sure!" You know, working our asses off, and trying to find the thing. Well, to make a long story short, we didn't find it. We just couldn't get all the rocks out. . . . So we had to give up. We put all the rocks back in and went to work on the Surveyor. . . .

And then it was time to pack up and get off the lunar surface. . . . I was putting the last rocks in the rock box, emptying the tool carrier. And, clunk, out on the top, there it is. So I picked it up and gave it the old heave. It probably went two or three hundred yards. So, it's up there somewhere. And I've always figured two million years from now, some archaeologist is going to be up there tromping around. . . . He's going to find this little blivet that there's no documentation on, and he's not going to know what it is.

—PETE CONRAD

What you remember of the lunar surface is more having looked at the picture over and over again, and looked and studied it. I remember more about what the moon's about because of the photographs that we took that I look back on, that give me a better recollection of it than is up there (points to his head).

—BUZZ ALDRIN

You know, a person when he comes down from that flight, ought to sit down with a tape recorder and just talk, just go off by yourself somewhere and talk.

—STU ROOSA

My most cherished memory of having been there is to look back at the Earth. Because it's just such a unique—it's alive, it's beautiful, as I said, it's your identity with reality. And you don't want to lose that.

—GENE CERNAN

I think what you do is you fall into a routine of either the twenty-minute or the forty-minute speech and questions and answers. And you never get back to really thinking about it.

—STU ROOSA

I wish I could go back to the moon, step out, everybody would go do whatever they had to do, and I could look, and talk about it, you know, record it. Look up-sun. Look for these things as an artist would. . . .

—ALAN BEAN

Money should have been spent to find one guy—and I don't know how you pick them, but you need someone—you need a Hemingway that can capture the feeling and describe, to say, "I saw this. And I saw that. And I saw this." That would have been worth the price of Apollo.

—KEN MATTINGLY

Things inevitably recede. I think when you're fifty-eight years old, things are not nearly as vital and important to you as they are when you're thirty-eight. . . . The sharp edges get kind of rounded off, and things that were of vital importance are of some importance. And an experience of twenty-something years ago does lose a little of its vitality, I think, just through the passage of time, through nothing else. But, fundamentally, you know, as I said in Carrying the Fire, I have done things and been places you simply would not believe, and I keep that inside me. And that's nice, and I like that. I feel very lucky. If I had to sum it all up at this stage in my life, in one word, I'd just say "lucky."

—MIKE COLLINS

Once in a while I can say to myself, Hey, I'm living on borrowed time. You know, my life could've ended on April 13, 1970. And consequently, I'm still here. So, nothing but pluses.

—JIM LOVELL

Buzz Aldrin on the Sea of Tranquillity.

It's something you can't avoid. Every time the moon comes up, why, it's there. But I don't think it reminds me of Apollo every time I look at the moon *(laughs)*. I don't think that happens. . . . I think it would be honest to say that I'm reminded of it enough by *other* people that I don't *have* to think about it myself.

—NEIL ARMSTRONG

When I look at the moon, I can immediately visualize myself flying over it. . . . I can visualize the landing spot.

—AL WORDEN

That's the other thing drives people crazy, they know I'm *lying*. They say, "Don't you go out and look at the moon?" And I honestly don't.

—PETE CONRAD

I don't even think about it. But when you look at the moon, sometimes it's hard to realize we were really that close to it. . . .

—FRANK BORMAN

I tell people, it *catches your eye,* more than it otherwise would.

—JACK SCHMITT

To me, the moon was going to be a lot more interesting than it turned out to be. The moon is not *2001,* sharp corners, and all that; it's a pulverized pile of dirty beach sand. . . . The poets have hated my guts—[I've gotten] hate mail from poets. "Dirty beach sand?!" After the flight, they said, "God, what kind of an imagination have *you* got?" Well, that's what it looked like! I was being honest. . . . When other people say, "Oh, what a beautiful moon," I'll think to myself, Heh, if they could just see it up close. It's only beautiful because it's a long ways off; you can't see that it really isn't a very interesting place.

—BILL ANDERS

I look at it very three-dimensionally. I mean, I can look beyond the moon. I can *see* the three dimensions. I think, prior to going there, and I think most people just look at it as something flat out there in the night sky. But I look at it as a sphere. I can *feel* the depth of it. I *know* that it is in inner space. I mean, I know that it's not at the end of that blackness.

—GENE CERNAN

Regardless of what Pete Conrad says, I look at the moon all the time. And I say those things: "I was there. . . ." One thing, you look at it, and it seems like it was so very, very real and such a short time ago. The other thing, you look, and you think, Huh, . . . did that re-

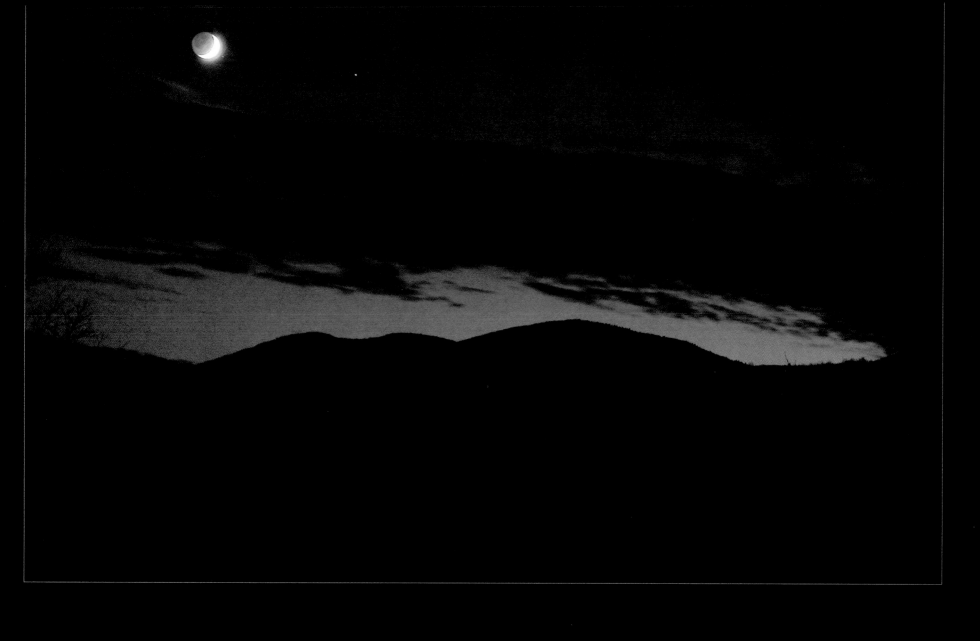

The moon, partially illuminated by Earthshine, and Venus over the mountains of Vermont.

THE SPIRIT OF APOLLO

Our destiny, at that time, was going to the moon. In fact, probably one of the clearest definitions of an objective or a destiny that mankind has ever experienced has been "man, moon, 1970." How could it be any clearer than that?

I've always believed in exploration. . . . Exploration is the greatest adventure. And exploration is why we're no longer huddled up in caves. Or no longer huddled up on the eastern seaboard, in thirteen colonies. Or why we carved this tremendous nation out of a wilderness. . . . This spirit that took us to the moon is the same spirit that moved our forefathers west across the country. And as they carried the flag west, why, we carried it on to the moon.

—STU ROOSA

Apollo didn't die. It accomplished exactly what it set out to do. It was a program of limited scope. . . . That's one of the strengths of our society. You know, you do it, and then you say, "Ho hum, gee whiz, what's next?" It's a great strength. The exceptional becomes ordinary and then forgotten. It's a strength, because we're always looking for the next thing. We don't accept progress as definitive, or as final. So we say, Let's go on; let's get on with it.

—FRANK BORMAN

I've always been a believer in pushing out the frontier. . . . It was kind of nice that the moon got in the way.

—ALAN SHEPARD

It was an exploration mission. We took the human intellect and the human vision, the human mind, 240,000 miles away from its home base. *That* was the importance. Whether we found a rock there or not was of no importance.

—FRANK BORMAN

Looking homeward from the orbiting Apollo 15 command module, Endeavour

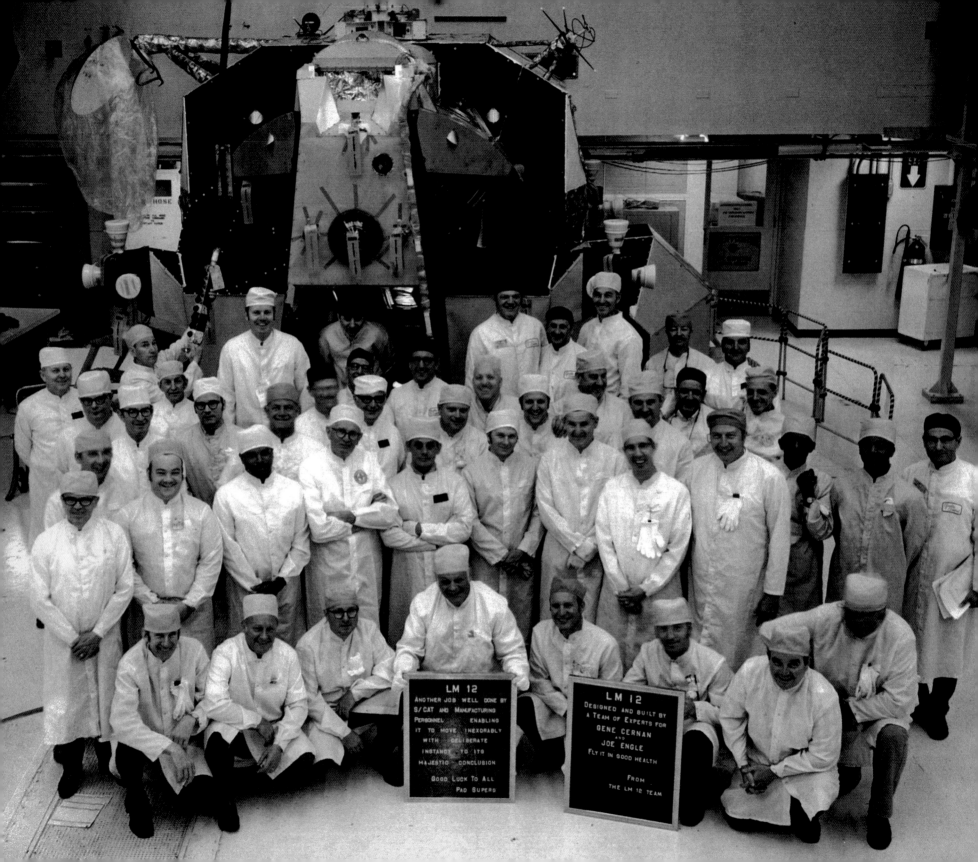

LM 12
ANOTHER JOB WELL DONE BY
S/CAT AND MANUFACTURING
PERSONNEL ENABLING
IT TO MOVE INEXORABLY
WITH DELIBERATE
INSTANCE TO ITS
MAJESTIC CONCLUSION

GOOD LUCK TO ALL
PAD SUPERS

LM 12
DESIGNED AND BUILT BY
A TEAM OF EXPERTS FOR
GENE CERNAN
AND
JOE ENGLE
FLY IT IN GOOD HEALTH

FROM
THE LM 12 TEAM

One of the other beauties of the Apollo program was the involvement of the people, which you can only appreciate by having seen it. . . . The Apollo program was such a nice environment in which people had personal involvement, personal dedication. And that's one reason it worked. It wasn't just 400,000 people doing their eight-to-five job, it was 400,000 people involved in a manned expedition.

—DAVE SCOTT

I probably had a dim glimmer of appreciation at the time. But as I get older, I can appreciate what a magnificent series of events came together to allow that to happen—and it's people. . . . It was being part of a team that was dedicated to something that transcended individual aspirations. . . .

Human nature being what it is, people have to have something they can put [their arms around]. There has to be a Rambo. There has to be a symbol. The folks who made the trip were the symbol. And that's unfortunate, because it takes away from the real elegance. You can take any one of those folks out for a beer, and they'll tell you stories about what their contribution was—and it's all fascinating. . . . Those people never got their names in print.

That's what Apollo was. It was thousands of people who were willing to work day and night. Not being compensated, not compared to what they put in. No one person [could] do it. You can't imagine what that's like compared to an everyday experience. . . . That's why we're so screwed up—we don't have many experiences like that. I don't think people can understand the exhilarating feeling of being part of that kind of endeavor—unless you've been there. The extraordinary things that can be accomplished when you [dedicate] yourself to the end result instead of your personal temporal consideration. Part of what comes out of that is the belief we can do anything in the world if we really honestly decide to go do it.

—KEN MATTINGLY

Nobody today understands the complexity, and how well the system really, really worked. . . . But, God, it was complicated. . . .

Look at the system we had, and how well it worked. Not one single LM failure. Not one single backpack failure. Not one single rover failure. . . . That's magnificent! *Nobody* builds hardware and does that. Airplanes don't do that. *Nothin'* works that well. I never flew an airplane that had that kind of record. Any of 'em! They *always* had problems. You know? Some of 'em crashed, right? None of that stuff crashed. I mean, it was a great system. . . . Great system, great people. . . . It was that way throughout the program. Everybody. The engineers were the best engineers in the world. The flight-ops guys were the best flight-ops guys in the world. That's the beauty of the program, I think, from the astronaut point of view. The association with the best people in the world.

—DAVE SCOTT

Workers at Grumman Aerospace in Bethpage, New York, pose with the lunar module they built for Apollo 17.

You may not believe this. But the true pilot doesn't do it for fame. Or fortune, either, for that matter. He really doesn't. Because it's his ass he's playing with. . . . The point is, you don't do it for the fame and fortune. Sure, it's nice to be recognized—*for what you've done.* Whether it's a suborbital flight, it's an orbital flight, whether it's a preliminary flight in the Apollo program. I think if you take everybody apart and you get right to the basis of why they were there, why they were training, and why they were risking their lives, it's that they're *damn proud* of what they've done.

—ALAN SHEPARD

Apollo veterans appear at the Experimental Aircraft Association's annual gathering in Oshkosh, Wisconsin, July 1994. From left: Charlie Duke, Al Worden, Stu Roosa, Dick Gordon, Pete Conrad, Mike Collins, Buzz Aldrin, Neil Armstrong, Gene Cernan, Apollo 9's Jim McDivitt, Jim Lovell, Bill Anders, Frank Borman, and Apollo 7's Walt Cunningham and Wally Schirra. Facing page, insets: Dick Gordon and Alan Bean, 1999; Fred Haise and Ken Mattingly, 2000; Neil Armstrong and Gene Cernan, 2002.

I thought about it when we left the surface. . . . I knew it would be a long time. I just felt it might very well be a generation before we get back to the moon. I'm probably going to be proven to be right.

—GENE CERNAN

Gene Cernan's photograph of Jack Schmitt near the end of Apollo's final moonwalk.

Everything else being equal, you could live on the moon. Self-sufficient. The moon could be as much of an outpost for human-kind as Plymouth was, at one time. . . . And in so doing, you also can develop a lunar economy that will be self-sufficient, that won't require subsidy, and indeed can profit from exports to space and exports to Earth. All that's very realistic now. I'm getting more and more confident that that's possible. I don't know who's going to do it, but it's possible.

—JACK SCHMITT

We were a base. . . . I mean, if we'd had more oxygen and more food and more water in the LM, we could've stayed there. Nothing wore out. . . . The rover's still good. You could go out there and start it up right now. So, we had a lunar base established. We just didn't have the supplies. If we'd been resupplied, we could've stayed there. So we—the world, the USA—know how to establish an outpost on the moon. . . . And if you ask me, What kind of space-suit would you want to wear on the moon, now that we've got all these fancy things? I'd go back to the museum and get my old one. Because it was marvelous. And if you asked me, What kind of rover do you want on the moon? I'd go get the one I had before—it was terrific! Yeah, I'd update it now with the better microprocessors and lighter material and all that. But it's still going to have wheels on it. . . . So I don't think we lost what a lot of people might think we lost by terminating Apollo. We developed that system. We know how to do it.

We've learned how to explore the moon. We've learned how to live on the moon. We've learned a little bit about the geology of the moon. Now you go back and really do it. Your explorers have gone to the West and they've blazed a trail. Now you send the pioneers out there! . . . And that's what people used to do in the Old West. They had a fort. They'd go to the fort, pioneers with their covered wagons. They'd stage out of that and then go build their little farms and things. Great analogy. Now you're getting to something that people can hold on to. It's tangible. So it says, Apollo had a reason. It taught us how to go into space and set up our first outpost. Now we go set it up. Now we go to Mars.

I think the moon's got a lot of things up there to tell us that we need to go find out. That's how I feel about it. We ought to go back and do it again sometime. Because it sure is an interesting place.

—JOHN YOUNG

When we go to Mars, it's not going to be the same. When we went to the moon, we went out and saw this great rocket go Whoosh! Boom! It took [three] days to get there. If you wanted to remember, you could go look at the moon. You can see a TV picture. People say, "I can remember that they landed. It was an amazing experi-ence. I saw where they went and where they came back." It's all within two weeks. When some cat goes to Mars, you might remem-ber the launch—and then next year, they're gonna talk to you from Mars. "Who are you?" They're gonna have to look up the picture to see what you look like. And a year later [when they come back from Mars], "Where have you been? Were you on travel? I've missed see-ing you. . . ." Humans don't have that long an attention span.

—KEN MATTINGLY

I have great ideas about how to bring about lunar exploration and space transportation, how to bring about Mars exploration. . . . You do what you're good at, and I realized that I'm good at that.

—BUZZ ALDRIN

Wanted to leave a Bible on the moon. Wanted to leave something of everything we had that was meaningful. Somebody's going to go back there one of these days, right? We left a bit of everything . . . flora, fauna, Bible, coins, flag, checklist. Can somebody reconstruct who we were? Yes. Because we consciously thought of that before we went. . . . I don't know whether the other guys did something or not, but it seemed to me that we should leave a trail. Leave a trail of who we were and what we were, for whomever someday. And you figure that stuff's been, you know, a billion years without getting disturbed. This'll be there for a billion years. Somebody will go back and pick it up. Why not leave them a story?

To me, space is nothing more than a continuum. . . . The fact of visually seeing [the Earth as a ball] is pretty astounding. . . . It's an unusual perspective that ultimately will become old hat. . . . When the day comes, and it will come, when people are going into space like you do today on an airliner, that won't be no big deal. The next big deal, and that *will* be profound to me, is the first guy that lights out for the next star system, and our sun goes away. When you get out in that void, now, that won't be the same effect of the scenery being exotic. It'll be the fact [that] all the scenery is the same, in every direction you look. And that has got to be some trepidation for whoever takes that step for the first time. See, we ain't left the solar system. You can look out the window, even on the way to the moon, you can see the sun, you can see the moon, you can see the Earth. We ain't really left. Whoever goes to the next star, or wherever, someday—they can be clear in their mind, they've left.

—FRED HAISE

Oh, but this [why you send people] is so clear. You can't explore without the perception and the judgment and the awareness and the intuitive nature of man. You just can't do that. Man is going to explore the universe, and pioneer, and settle the universe. There's no question; it's just when. You don't have to do it this year or next year, but it'll get done.

—DAVE SCOTT

This page: *Earth and moon, photographed from a distance of 7.25 million miles by the unmanned spacecraft Voyager 2. Opposite: Stars and clouds of dust and glowing gas in the Large Magellanic Cloud, 160,000 light years distant, photographed by the Hubble Space Telescope.*

Apollo is still to me a great adventure. The most adventurous thing that I've ever done, in the human sense. Not the most adventurous, because I've seen God's supernatural power in my life, and that's more exciting than going to the moon. But in the human sense of the adventures that we can have in technology, Apollo is a great adventure. It's a great step of Man. I think it will go down in history as a reach out to the inquisitiveness, the sense of exploration, the sense of adventure, the quest for knowledge that Man has within us, and it will always be that to me.

—CHARLIE DUKE

Apollo was such a beautiful, awesome, magnificent program. Probably never repeated in anybody's life, whether it's a Mars landing or whatever it is. . . . As astronauts we got the brass ring and got to fly. But all the engineers and everybody else in NASA felt the same way. There was a spirit of the program. And it was a beautiful thing. And I seriously doubt if it could ever be repeated under any leadership or any mission. I think it'd be very difficult to get that feeling back again. Because it was—well, I think *National Geographic* said it all: "Mankind's Greatest Adventure."

—STU ROOSA

Gene Cernan with the Stars and Stripes at the valley of Taurus-Littrow.

OPENING SPREAD CAPTIONS

ACKNOWLEDGMENTS

We are grateful to the following for help in obtaining photographs: Mike Gentry, Susan Phipps, and Adam Caballero at NASA's Johnson Space Center; Mark Gray of Spacecraft Films; Kipp Teague, creator of the Apollo Image Gallery (www .apolloarchive.com/apollo_gallery.html); Jeannine Geiger at the Air Force Flight Test Center History Office, Edwards AFB; Daniel Gruenbaum, Astronaut Hall of Fame; Mary Weeks.

PHOTO CREDITS

All images NASA unless otherwise indicated.
Courtesy Jim Lovell—p. 2 left.
U.S. Military Academy—p. 2 bottom left.
U.S. Naval Academy—p. 2 bottom center, right.
Courtesy Jack Roosa—p. 3.
Courtesy Alan Bean—p. 4 top.
Courtesy Tom Stafford—p. 4 bottom.
Air Force Flight Test Center History Office—pp. 5, 6 bottom, 7.
Courtesy Jan Evans—p. 6 top.
Mark Gray—p. 157.
Andrew Chaikin—p. 158 (both), p. 181, pp. 188-89 (all).
Grumman History Office—p. 186.
Hubble Space Telescope Science Institute—p. 195.

High-quality prints of Apollo photographs may be obtained by visiting
www.andrewchaikin.com.

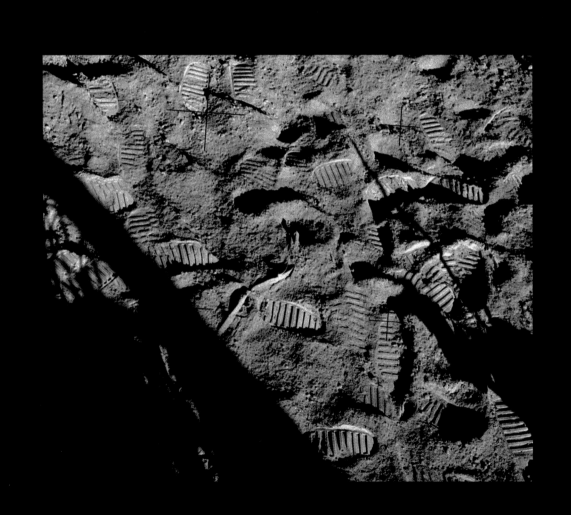